AF689606

HEALING GROUND, LIVING VALUES
Stanley Center for Peace and Security

An Ecotone Publishing Book/2024

Ecotone Publishing is an Imprint of the International Living Future Institute

International Living Future Institute
PO Box 5874
Portland, OR 97228-5874

Author: Justin R. Wolf
Book and Cover Design: Johanna Björk, softfirm

Library of Congress Control Number: 2023951475
Library of Congress Cataloging-in Publication Data

ISBN-13: 978-1-7362129-5-0

FSC www.fsc.org MIX Paper | Supporting responsible forestry FSC® C016245

First Edition
Printed in Canada on FSC-certified paper, processed Chlorine-Free, with vegetable-based ink.

HEALING GROUND, LIVING VALUES

Stanley Center for Peace and Security

LIVING BUILDING CHALLENGE

FOREWORD 3

INTRODUCTION 4

ACKNOWLEDGEMENTS 6

AUTHOR PROFILE 7

PART I:
AN EVOLVING CITY
The Pearl of the Mississippi 10
The Stanley Center 25

PART II:
FINDING A NEW HOME
Getting Started 38
Shared Values 41
Sustainable Potential 43
Analyzing the Musser Library 47
Creating Ownership 49
Ground Healing 52

PART III:
FORGING A NEW HOME
Petals, Imperatives, and Performance 58

THE PLACE PETAL
Referencing Place and History 65
Meeting of the (Biophilic) Minds 72
A Courtyard for Food and Community 74
Good Neighbors 77
A Walkable Neighborhood 79

THE WATER PETAL
Honoring the Watershed 85
Collection, Storage, and Responsible Use 89
Handprinting with MCSA 93

THE ENERGY PETAL
A Roof That Isn't Big Enough 97
Savings and Recovery 100
Designing a Passive System 104
Resiliency Strategy 107

THE HEALTH + HAPPINESS PETAL
Discovering Purpose 111
Designing a Living Room 112
Uncovering Light and Landscape 114
Activating Every Corner 118

THE MATERIALS PETAL
The Most Challenging Petal 122
Red List Compliance 125
All In the Family 131
Responsible Sourcing and Salvaging 133

THE EQUITY PETAL
Becoming a Just Organization 143
Achieving Accessibility 145
Sustainable Action and Social Justice 148

THE BEAUTY PETAL
Where the Petals Converge 154
Transformation Through Nature 157
Transformation Through History 158

PART IV:
A HIGHER CALLING
Reflections from the Stanley Center 164
Reflections from Neumann Monson 167
A Lasting, Living Impact 168

APPENDIX:
Awards 170
Photographs and Illustrations 170
More Books from Ecotone Publishing 172
International Living Future Institute 174

FOREWORD

BY ROBERT C. ORR
Professor and Dean, School of Public Policy, University of Maryland
Special Adviser to the UN Secretary-General on Climate Change

At the heart of the climate crisis is a stark reality: our planet's health is deteriorating at an alarming rate, threatening ecosystems, human livelihoods, and the very fabric of life on Earth. Climate change has exacerbated food, health, energy, and economic crises in every corner of the planet with a compounding effect on inequity, division, and conflict. But ambitious climate action—done right—will lead to a world better than we found it.

The climate crisis, a manifestation of decades of unsustainable development, exploitation of natural resources, and a lack of sufficient global action, demands immediate and bold responses. Through its living building, the Stanley Center exemplifies a commitment to sustainability that embodies the urgent need for increased climate ambition and collaborative action across all levels of society. With its design and construction partners, the center demonstrates how sustainability and innovation can merge to create solutions that are environmentally sound, economically viable, and socially equitable.

The Living Building Challenge, arguably the most rigorous sustainability certification for buildings, sets the bar high. It demands more than mere energy efficiency; it requires that buildings produce more energy than they consume, harvest rainwater on-site, provide occupants with access to daylight and healthy air, and use materials that are safe for all living creatures in the long term.

The Stanley Center's decision to embrace this challenge is a testament to its vision in the fight against climate change and underscores the importance of the built environment in mitigating this global crisis. As this book reveals, the center's living building demonstrates that our buildings, often criticized for their significant carbon footprint, can instead become part of the solution—a vital insight for architects, developers, policymakers and citizens.

Addressing the climate crisis demands that we envision a future that goes beyond merely reducing emissions to one where our activities positively contribute to the health of the planet. The Stanley Center's home is not just about creating a sustainable building; it is about reimagining our relationship with the natural world and upholding safe and just solutions that redress and repair systemic harms.

As global temperatures move toward the Paris Agreement target of a maximum 1.5°C rise, the window for meaningful action is closing rapidly. Efforts like the Stanley Center's living building are crucial not only for their immediate environmental benefits but also for their ability to inspire others. They demonstrate that ambitious, forward-thinking projects are both feasible and necessary.

This book is more than a mere chronicle of a sustainable building project. It is a call to arms—a plea for increased ambition, collaboration, and decisive action. The Stanley Center's new home is a shining example of what is possible when we commit to future generations and acknowledge that each of us has a role to play in the life of our planet.

INTRODUCTION

"When you think about the Stanley Center's mission and vision, it became clear to everyone that the Living Building Challenge was the right fit and reflected what we are trying to achieve as an organization."

WARREN HEIDBREDER
Member of Governance, Stanley Center for Peace and Security

A Living Building is regenerative and self-sufficient. It gives more than it takes. It is of its place, fully embedded in the context of its surroundings, native bioregion, community, and culture. There is no off-the-shelf version of a Living Building one can purchase, and no design software that can specify a Living Building on command. From the program's earliest renditions up to present day, the International Living Future Institute's Living Building Challenge embodies the highest ideals and best-in-class practices that come from collaboration, transparency, and systems thinking. When individuals consider any Living Building, whether as a designer, builder, manufacturer, supplier, occupant, third party verifier, or even casual observer, they must be conscious of the building's place. This consideration, above all, is paramount. So, what kind of place are we talking about here?

This story begins in a small river town in the American Midwest, fronting a wide and curving bank along the Mississippi River. Muscatine, Iowa, is a place of industry and trade, of abundant resources and rich knowledge. In the annals of history, the city has played an important and often decisive role in advancing progresses in lumber manufacturing, civil rights and school integration, the transcontinental railroad, engineering, furniture, fashion, journalism, philanthropy, as well as social and environmental justice. Situated just thirty miles downriver from the Quad Cities (a region of five cities located in Iowa and Illinois: Davenport and Bettendorf, in southeast Iowa, and Rock Island, Moline, and East Moline, in northwest Illinois), where the river bends westward, Muscatine is one of just a few river towns of some consequence where the Mississippi flows east to west. In his 1883 memoir Life on the Mississippi, Mark Twain, who lived and worked in Muscatine for a brief spell in his youth, recalls, "And I remember Muscatine — still more pleasantly — for its summer sunsets. I have never seen any, on either side of the ocean, that equaled them."

No other city, town, village, or territory in the world claims the name Muscatine. In both name and history, the city is one of a kind. And now, a future Living Building resides in downtown Muscatine, slated to be the first of its kind for Iowa and the surrounding region.

The Stanley Center for Peace and Security is a non-profit foundation devoted to facilitating a wide range of policy changes and governance practices that address global challenges tied to climate change, nuclear weapons, and mass violence and atrocities. The organization was founded in Muscatine, in 1956, by Max and Betty Stanley, a devoted and enterprising couple who laid down roots and started a family in Muscatine, who built industry and enormous wealth in Muscatine, and who ultimately, and consequently, created a mission-driven, values-based organization dedicated to the advancement of novel ideas and education for solving the world's greatest global challenges, right in Muscatine.

"The old Musser Library was a good space for patrons, with some minor amenities. The building was conducive to being a good library. Open space, lots of natural light, but we didn't really have the space that enabled collaboration. It really impacted our programming."

ROBERT (BOBBY) FIEDLER
Director, Musser Public Library

"The message this Living Building sends is that this is possible and doable on a very practical level. At the moment, you pay a premium to achieve this level of performance. But that premium can be reduced, and you get to that point by actually doing more projects like this, by opening production lines and encouraging change in the manufacturing sectors and changing people's thinking when it comes to design and construction practices."

KEITH PORTER
CEO and President, Stanley Center for Peace and Security

The Stanley Center and its staff of twenty-five moved into their new headquarters in March 2023, while the finishes were still being applied and roof-mounted solar panels were being installed. Adapted from a 1970s-era public library that had been sitting vacant since 2017, the building itself was architecturally bland. Some residents described its bulky form as "intimidating" and "fortress-like." While inherently civic in nature, and fondly remembered in passing by some, the physical building would require serious changes to become a suitable candidate for reuse. But it had good bones and a strong foundation.

"The future of the old Musser Public Library building remains uncertain" reads a June 2018 article in the Muscatine Journal. In the same article, the city's communications manager Kevin Jenison is quoted, "It would take someone with the right vision to redo that space."

Indeed, the right vision came about. That vision begins, obviously, with the staff and leadership of the Stanley Center. It is also the doing of Neumann Monson Architects, Graham Construction, HNI Corporation, the City of Muscatine, and several other key stakeholders whose contributions and capacity for foresight will be covered in further chapters.

This future Living Building in Muscatine is a story of adaptive reuse. More so, it is a story of an organization (and several individuals) with deep roots in its community doubling down and making good on its commitment. The new headquarters of the Stanley Center for Peace and Security — situated on a sloping intersection at the corner of Third Street and Iowa Avenue, in the shadow of some of the most crucial milestones in American industrial and civil rights history — is a place designed to embody the organization's founding vision to promote "peace with freedom and justice." It is a place that stands for reparations for those Indigenous peoples who once made their homes and livelihoods on this land, and for healing the wounds inflicted upon that land since settlers arrived in the age of manifest destiny.

The Stanley Center is integrated within its community and its home includes space dedicated to the Center's programming in Muscatine. This approach goes to the Center's founding charter of being an organization that not only advocates for global policy change and promotes peace and freedom on the world stage but is likewise invested in bettering the very community it has called home for over half a century.

"It is possible to create a building that is forward thinking and values driven," says Keith Porter, CEO and President of the Stanley Center. "And this building is a physical manifestation of the history and values of the Stanley Center."

ACKNOWLEDGEMENTS

Healing Ground, Living Values could not exist without the dedicated and tireless efforts of the Stanley Center for Peace and Security and the organization's governance. To Keith Porter, Mark Seaman, Jennifer Smyser, Krista Regennitter, Dane Lovell, Patty Papke, Amy Bakke, Cassie Batzkiel, Alex Leites, Catherine Elliott, Warren Heidbreder, and the entire team, thank you for your good humor and transparency. Good values attract good people, and your collective work is a gift to humankind. If ever I should find myself despondent or cynical over the state of the world, I will be well served to consider the example of your organization's founders Max and Betty Stanley, as well as the mission and achievements of the Stanley Center, past, present, and future.

To the Neumann Monson project team of Tim Schroeder, Khalid Khan, Lyndley Kent, Sally Obernolte, Brandon Fettes, Jillian Riggan and Denise Schmitz, my gratitude is boundless. Congratulations in spades for this distinct design achievement, and for hurdling each challenge throughout this journey – during a pandemic no less – with grace, creativity, and diligence. You raise the standard for our industry!

High praise and thanks are owed to the leadership and core team members of Graham Construction, HNI, Design Engineers, Integrated Eco Strategy, Environmental Consulting & Technology, Biohabitats, Raker Rhodes, Muscatine Center for Social Action, Musser Library, Ideal Energy, Certified Wood Products, M.L. Johnson Woodworks, and of course the City of Muscatine. The following individuals (listed alphabetically) were indispensable guides as I pieced together the interrelated histories of the Stanley Center, the Stanley family legacy, and so much more, not to mention the complex technical facets and design concerns of the building itself: Robert Albee, Brad Bark, Lisa Brunie-McDermott, Darrow Center, Rachelle Chase, Dan Clark, Deana Dartt, Robert Fiedler, Paul Finkelman, Dan Haugen, Diane Haugen, Brad Hill, Jonathan Lawrence, Bryanna Johnson, Matthew Johnson, Tim Lentz, Brian Mrozek, Pete Munoz, Mary Palmer, Justin Paterson, Kent Sissel, Tyler Stone, Amy Van Beek, Troy Van Beek, Brenna Wissman, Shayla Woodhouse, and Ginger Zenor. Thank you all for your time and candor. I am in your debt and I am wiser for having crossed paths with you.

Michael Berrisford, your encouragement and steady editorial hand throughout this process have been invaluable. I'm glad to know you, Mike. Thanks must also go to Fred McLennan, Johanna Björk, and the entire Ecotone Publishing team for your vision and leadership. Thank you as well to Kira Gould and Pete Smerud, who independently facilitated my introduction to Ecotone back in early 2022 and recommended my services. I must be doing something right in this life to have earned your confidence.

Thank you to my family, my heart, Danielle, Jonah, and Violet. You give me strength. Thank you, Mom and Dad, for being my heroes. And finally, I offer gratitude for the examples set by my grandfather and uncle, Bob and Bobby. This book is written in your memory.

JUSTIN R. WOLF

JUSTIN R. WOLF

Justin R. Wolf is an independent writer, editor, and content strategist for the AEC industry. His byline regularly appears in *Green Building Advisor*, where he is a contributing editor and writes about energy policy, green building news, and climate action milestones. His articles, essays, and op-eds have also been featured in the pages of *Fine Homebuilding, Metropolis, Architectural Record, ENTER* (the magazine of AIA Minnesota), *Common Edge, Arch Daily, Fast Company, Frank Lloyd Wright Quarterly*, and other publications.

Complementing the daily grind of a freelance journalist, Justin also operates a small advisory, ghostwriting, and content strategy business that partners with AEC firms to craft stories and pursue publishing opportunities for established and emerging thought leaders in the design, engineering, construction, and manufacturing fields. He proudly counts many of the industry's most innovative and progressive practices among his client roster, past and present.

In his past lives, Justin worked as an educator, teaching history and environmental science; an independent art writer and public relations consultant for several commercial art galleries; and for more than a decade performing in-house media relations, marketing, and communications services on behalf of international architecture and integrated design firms. At the onset of the Covid-19 pandemic, Justin returned to writing full time and hasn't looked back since.

Healing Ground, Living Values is Justin's first contribution to the Living Building Challenge Series. He lives in southern Maine with his wife and two children.

PART I

An Evolving City

WELCOME TO PEARL CITY

The Mississippi River waters slow to a steady crawl and shimmer in the daylight as they flow past Muscatine and the forested bluffs of Illinois City on the opposite side. This occurs near the terminus of the river's long westerly bend before turning south once again, en route to Fort Madison, Keokuk, and Hannibal, Missouri. Gazing westward, plumes of murky exhaust emanate from a power plant in the distance. Along Riverside Park, bisected from the main drive by tracks carrying rail freight, the towering bronze monument "Mississippi Harvest" depicts a shell miner holding aloft a pair of clamming forks three times his height. The plaque affixed to its base reads: "Dedicated to the men and women who made Muscatine the pearl button capital of the world." The Pearl City Station stands sentry at the edges. From this vantage point, Muscatine is a quiet and beautiful place, even as freight trains rumble through and the stain of industrial pollution rears its head. It's a town with "class," as described by the English writer Jonathan Raban, upon stopping ashore during his two-thousand-mile journey down river in 1979. "A nineteenth-century river town that was in remarkably good working order," Raban wrote.

Muscatine is a city that is ever evolving, with a history and charm that beckons exploration on foot.

Downtown Muscatine slopes upward from the river's edge. Ascending Iowa Avenue and entering the city's commercial historic district, an elegant charcoal-gray building stands firmly at the avenue's intersection with Third Street. The physical building itself doesn't so much stand out, but in many respects — amidst the industrial town's densely packed blocks and weathered buildings of brick and limestone — its design does.

The newly renovated two-story building at 304 Iowa Avenue is clad in vertical bands of sustainably sourced Accoya siding, charred to a darkened gray for durability. Facing Third Street, vertically oriented windows are punched into the façade in a barcode-like procession. The building's expansive entryway, facing the avenue, features a large canopy overhang supported by slender steel girders; solar panels are suspended above, and beneath them on the far end, twin cisterns sit side by side, inviting passersby to stop and ponder the presence of two large aluminum drums facing the sidewalk. Near the foot of the entrance, three stacked bricks are embedded into a raised buffer, marked with the years '1901', '1971', and '2021'. A circuitous path of bricks winds its way through the herringbone pattern of tiles along the curb, resembling a river run.

Inside, natural light pours into the double-height atrium through three pitched skylights. Reaching for that light are four Ficus Benjamina trees, standing in symmetrical formation and each one rooted beneath the foundation. Potted plants of Golden Pothos, peace lilies, palms, and Ficus frame a variety of lounge and common areas. Towards the back, two wall-mounted installations of stacked, irrigated planters pour forth with edible lettuces. Air plants rest in a bowl of unfinished pearl button "blanks," and hundreds more blanks compose intricately assembled backsplashes in the communal kitchen. Beyond the kitchen, where offices and some library stacks once resided, a substantial chunk of the building — nearly one-fourth of its existing footprint — has been removed to create space for an expansive open-air courtyard, where native Virginia Creeper climbs a two-story trellis mounted to the old library's existing exterior wall.

This striking composition of form and function is the new home of the Stanley Center for Peace and Security, an organization dedicated to catalyzing just and sustainable solutions to critical issues of peace and security, such as mitigating the impacts of global warming, avoiding the use of nuclear weapons, and preventing mass violence and atrocities. From the gutted remains of the city's 1970s-era rendition of their public library, which was "built like a tank," according to Neumann Monson's Tim Schroeder, something unique, regenerative, and restorative emerged. This building

is designed to generate 108 percent of its energy needs; it will capture and store enough water for all plumbing, irrigation, and potable water needs; and it features a generous urban agriculture program that provides fresh produce to building occupants and their community. But getting there wasn't easy for a host of reasons, one of which concerns the land itself.

While the building's footprint provided a sturdy bit of infrastructure on which to renovate and re-envision, that tank-like quality of the existing structure also largely concealed the site's unique ecological qualities and history.

The site slopes upward at a diagonal, from its southeast corner, at the intersection of Third and Iowa, to its northwest corner, where the building meets the adjacent alleyway, now canopied by a portion of the newly installed photovoltaic array. Traipsing the edges of the block, depending on one's vantage, this change in grade can appear dramatic, gradual, or illusory. The reason for this perception shift is an existing creek bed that flows beneath the building that was covered and redirected into a culvert in the late nineteenth century. Due to the culvert's underground piping, the building's footprint follows the ascending site at an unexpected angle, stepping up and back towards the far corner of the courtyard. While this subterranean feature obliged the project's designers to seek creative solutions, both inside and outside the building, the creek bed also informs the history and fates of buildings that preceded the 1970s-era library on this very land.

History is in plain sight in Muscatine, and one doesn't have to travel far past the Stanley Center's front curb to catch a glimpse. Across the street, a nearly two-century old apartment building was once a schoolhouse that happened to play a critical role in the history of American civil rights and school

integration. And just up the street, perched on a hill, is an unassuming brick house that was once the home of Alexander Clark, the civil rights icon whose own story and achievements were, until recently, largely ignored. On the opposite corner are the current offices of Stanley Consultants, the engineering firm that originally brought Max Stanley to Muscatine, in 1932, and was instrumental in electrifying large swaths of the Midwest in the 1930s and 1940s. Embedded in the soil of this river town on a bluff are tangible traces of its industrial past, most notably remnants of the freshwater mollusk shells that helped make the city famous more than a century ago.

Meeting the Living Building Challenge demands an iterative approach. In the case of the Stanley Center, that approach is not strictly limited to the building's design phases (pre-design, schematic design, design development, and so forth) to meet certain performance benchmarks. The iterative approach to realizing the Stanley Center's new building demanded a broader and more in-depth examination, not just of the organization's history and impacts, but of the city, its people, and the land. When we consider the achievements of the Stanley Center for Peace and Security through present day, and marvel at the singular accomplishment that is its new headquarters at 304 Iowa Avenue, we must also consider the long and storied historical arc of Muscatine, aka Pearl City.

According to Mark Seaman, vice president and director of communications at the Stanley Center, "If you tell the stories of Alexander Clark and the history of the Meskwaki people, there's an opportunity there to make that history come alive and convey the contributions of others that have come before us. It's critically important that we weave these and other stories into our story about the building. It would be a missed opportunity otherwise."

"When Illinois settlers began to pour in and take over, some of the Meskawki made agreements with some of the settlers. American history characterizes this as, 'The Meskwaki ceded this piece of land.' But these [were] gunpoint treaties, with the threatened extermination of your people by a huge military if you don't sign. In 1845, Meskwaki people were forced out of Iowa. The U.S. government took Meskwakis to southwest Kansas. We left with more than 1,000 people, and [12 years later] we came back [to Iowa] with a little more than 300. That was a genocidal event for us."

LEAH SLICK-DRISCOLL
Meskwaki Tribe, Winnebago Nation of Nebraska, Educator

WHERE IS MUSCATINE?

"The city of Muscatine derives its importance, in a commercial point, mainly from its peculiar location."

This is an excerpt from an article in *Ballou's Pictorial* (Boston), published on August 4, 1855. It continues: "It is situated on the extreme western bend of the Mississippi River, and is thus thrown into the interior, as it were, of the State. It is the landing point on the river for a large tract of country, and in this respect has the advantage over every other city in Iowa." For anyone unfamiliar with Muscatine, they might be intrigued by the inherent and complex contradictions described in this passage. The city lies along an "extreme western bend" at the state's eastern anterior yet is "thrown" into the interior due to the bend's dramatic push/pull. Where might such a place exist?

Muscatine is a city of roughly eighteen square miles located on a series of bluffs and hills hugging a west-south bend of the Mississippi River. The unique orientation of the bend gives the city approximately 260 degrees of riverfront. This expansive exposure speaks to the city's fortuitous and often complicated relationship with the mighty river, dating back to before European settlers and traders arrived at its shores.

The earliest chronicled people known to have lived on these lands and tilled its soils were the Iowa (aka Ioway) people, a Siouan tribe that grew maize, traded furs with the French, and adopted horses for hunting. The Iowa's domestic villages consisted mostly of teepees and lodges constructed from earth and bark, all designed toward achieving a natural equilibrium in both life and livelihood.

During the latter half of the seventeenth century, a combination of aggressive trading and hunting, colonial genocide, and tribal wars throughout the east and Great Lakes region forced many Indigenous tribes westward. Seeking refuge from both European and Native enemies, the cousin tribes of Meskwaki (aka Sac and Fox tribe of the Mississippi in Iowa) and Sauk peoples settled in the vicinity of the Quad Cities, just upriver from Muscatine. The two tribes maintained centralized yet largely separate villages near modern-day Rock Island, Illinois, at the confluence of the Mississippi and Rock Rivers. Adopting semi-nomadic lifestyles, together the Meskwaki and Sauk developed sophisticated systems of agriculture along the river's shores (predominantly corn, beans, and squash), hunted in the north and west during the winter months, and traded with other regional tribes and settlers during peace times. Local historian Daniel G. Clark refers to these villages as a "veritable supermarket," with abundant crops and "plentiful game." But with the arrival of white settlers came the all-too common practice of so-called "gunpoint treaties" and forced removal carried out in the name of colonialism.

The first U.S. settlement of Muscatine occurred in 1833, when Colonel George Davenport (the namesake of Davenport, Iowa), then stationed at Rock Island, dispatched representatives to establish a new trading post downriver. Davenport's dispatches, James Casey and John Vanater (aka Vanata or Vanatta), opened a supply depot for steamboats, called Casey's Woodpile (steamboats used wood for fuel). Consequently,

one of the earliest names for the informal settlement was Casey's Landing. The town's founding as a source for lumber (the first reported mill in town opened in 1838), even in a transient capacity, would prove prescient to the city's future as a place of industriousness and entrepreneurialism.

Once the county of "Musquitine" was established, in 1837, the town was named Bloomington — after Bloomington, Indiana, where Vanater, who laid out the town, once resided — and established as the county seat. (One year prior, a surveyor was sent to survey a town called Newburg, then with a population of just 71, but evidently the name didn't stick.) Another dozen years would pass until about 200 citizens of the town petitioned to have its name changed, owing to common confusion with other Bloomington's nationwide, and even with the town of Burlington, located 70 miles downriver. In the summer of 1849, the town changed its name once more, this time to Muscatine. While the derivations of that name have been debated — including as a variation on Musquakeen, an alternative name for Muscatine Island, as well as its possible links to the Meskwaki people — one indisputable fact is the enduring uniqueness of the name itself on the world stage. To this day, no other town on Earth shares its name.

With this small but growing community now established and officially incorporated by the mid-century point, with nearly 2,500 residents, Muscatine was well on its way to becoming a major center of commerce, trade, and manufacturing. Concurrent with this growth, the legacy of forced removal of the land's native inhabitants — the Meskawki were marched out of Iowa in 1845 — leaves a stain on the region's history that persists to this day.

View of the Mississippi River and South Muscatine, taken from Muscatine Heights, circa 1899

"Growing up here, my overall impression of Muscatine was of a really successful small industrial town that had an outsized impact. That sense of industriousness has always been a theme here in Muscatine."

ROBERT (BOBBY) FIEDLER
Director, Muscatine Public Library

"Clammers" working the river in search of mollusk shells, around the turn of the century

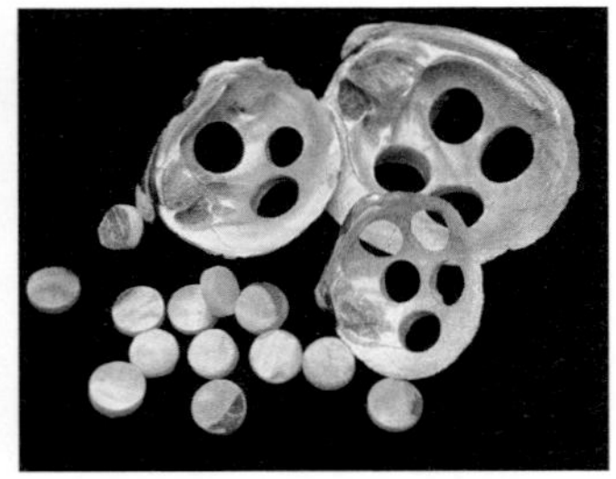

THE AGE OF EXPANSION

Not long after the official transition from Bloomington to Muscatine, expansion of the nation's railroads touched down in the city and set off a population boom that would nearly quadruple Muscatine's constituency and workforce over the next thirty years.

In many respects, beginning with the town's official incorporation at the turn of the decade, the 1850s were the most transformational period in Muscatine's history. The decade's happenings paved the way for new industries, new trade revenues, and new sources of civic pride that manifested themselves in wholly unexpected ways.

In 1855, an enterprising teenager from Hannibal, Missouri, named Samuel Clemens, arrived in Muscatine to join his brother and eldest sibling Orion, who was a co-owner and co-editor of the *Muscatine Journal*, the town's daily newspaper that is still in publication today. Samuel, who would of course soon thereafter achieve acclaim under the pen name Mark Twain, briefly worked for the *Journal* as a printer and reporter, remaining in town for "three months at the most," according to Daniel G. Clark, a local historian, *Journal* columnist, and former employee of the Stanley Center. (Despite Twain's very short stint in town, Muscatine boasts the Mark Twain Overlook, a bluff overlooking the Mississippi River and Illinois' shoreline just beyond, likely owing to Twain's flattering mention of the city in his memoir *Life on the Mississippi*.)

That same year, the Mississippi & Missouri Railroad held its inaugural passenger service, traveling from Davenport to Muscatine. Uneventful as the brief trip may have seemed at the time, this marked the very first passenger rail service west of the Mississippi River. Boom times were upon the town. "The number of inhabitants is about 6000, and is increasing rapidly," reads a later passage in the 1855 *Ballou's Pictorial* article. "The trade at this point is immense; far beyond any calculations that would be made by anyone not cognizant of the fact. New branches of business are being constantly introduced, and all classes of labor are in demand." This declaration was both timely and prescient. Muscatine, a largely blue-collar town of just over 20,000 residents, remains a manufacturing juggernaut to this day.

It was also in 1855 when two brothers named Richard and Peter Musser started their own lumber business in Muscatine, R. Musser & Company (later incorporated as the Musser Lumber Company, in 1881). Their nephew, Peter Miller Musser, who went by the initials P.M. to avoid confusion with his uncle, was recruited by Richard and Peter to move to Iowa from their home state of Pennsylvania and would eventually take over and grow the company from merely a supplier to a manufacturer of lumber wares. In 1871, the company opened a sawmill on the southern edge of town, colloquially known as "Musserville," and proceeded to churn out more than 11 million board feet of lumber.

In an era when timber barons reigned along the shores of the Mississippi, the name Musser carried serious weight. The Musser Lumber Company was a large shareholder in the Mississippi River Logging Company, which harvested and transported lumber principally from Minnesota, at the time the world's leading market in timber sourcing. (Interestingly, P.M.'s sole surviving child, Richard "Drew" Musser, would later co-own and operate the Pine Tree Lumber Company, in Little Falls, Minnesota, a business outfit he ran with his partner Charles A. Weyerhaeuser, whose father Frederick owned the Mississippi River Logging Company.) In the decade that followed the opening of Musserville, the company's lumber production grew to 50 million board feet annually, a planing mill was added for the production of finished boards, and distribution expanded as far north as North Dakota and as far west as Colorado.

Beyond lumber, the final decades of the nineteenth century saw a different and altogether unexpected form of wealth and trade come to Muscatine. In 1890, a German immigrant and button maker named John F. Boepple arrived in town after having discovered, through an American intermediary, an abundance of freshwater mussels and clams in nearby riverbeds. The shells' sturdiness and natural luster were qualities that Boepple knew how to harness. (Boepple was already an accomplished and versatile button maker, fashioning his wares from wood, bone, and horn, but oceanic mother-of-pearl were his top sellers; it was a newly imposed German tariff on imported shells that brought him to the United States.)

The distinct west-south bend in the river fronting Muscatine causes the waters to naturally slow down. Legend has it that Boepple noted this unusual bend on a map and hypothesized that mollusks had gathered by the millions along the beds of the low-current waters, waiting to collect the nutrients that over time yield opalescent pearls. His theory paid off. This discovery of freshwater shells quickly led to Muscatine becoming the epicenter of a thriving (and highly leveraged) mother-of-pearl button industry. It was also the industry that lent the city its new nickname: The Pearl of the Mississippi.

"The story of the pearl button is a national growth story, a national treasure story, and an environmental lesson," Terry Eagle, the Director of the National Pearl Button Museum, told *Atlas Obscura* in a 2019 interview.

In the ensuing years, one factory (opened by Boepple in 1895) soon gave way to dozens more. Shell miners, or "clammers," were employed by the thousands, including many local farmers who worked the rivers and creeks during their off season. Shells harvested from other riverbeds and marshlands along the Mississippi were shipped to Muscatine to be processed and sawed into rough-hewn, coin-shaped "blanks." By 1905, the industry hit peak production, with over 1.5 billion pearl buttons being produced annually, accounting for one-third of the world's pearl button output. At one point, approximately 50 percent of the town's workforce was employed in one form or another by the pearl button industry, and the city of Muscatine had rightful claim to the title of pearl button capital of the world. But in the age of unencumbered industrial growth, such superlatives always came at a cost. Indeed, the once relatively pristine waters of the Mississippi contained an estimated 300 different species of mollusks before Boepple's arrival. Less than two decades in, that figure had been reduced by roughly a third.

The pearl button industry was of course not long for this world, at least on a scale that once put Muscatine's operation on par with other industry cities like Pittsburgh and Detroit (albeit with a far smaller product). But long before consumer plastics were introduced and popular fashion trends became defined by convenience and frugality, Boepple's own story came to a tragic yet oddly poetic end. In the early days of 1912, while wading in the shallow waters of the river, he sliced his heel on a sharp clam shell and contracted a blood infection that claimed his life days later. And what possibly possessed Boepple to enter the frigid waters on that January day, presumably barefoot? One version of history — and a rather tall tale at that — has it that he was *returning* mollusk shells to the river, a small yet profound gesture (allegedly) of regeneration that just happened to cost him his life.

From lumber to buttons, each of these defining enterprises in Muscatine up to this point were predicated on extraction. Nature's bounty produced great wealth for the region and established the city as an industrial force. This chapter in the city's history also stood in stark contrast to how the land, river, and bluffs were harnessed and harvested by the Meskwaki, long before the days of Casey's Landing and Musserville. In due fairness, P.M. Musser would eventually give back to his community in significant ways that, in some respects, acknowledge the incredible generational wealth his family built from the literal roots of the region. (These contributions are explored further in the chapter "Third and Iowa.")

Another name that holds sway in Muscatine is Stanley. Claude Maxwell (Max) and Elizabeth (Betty) Stanley likewise built enormous wealth here and created businesses that further established Muscatine as a center of industry and industriousness. Max, a civil engineer by trade and educated at the University of Iowa, bought into the engineering firm Young & Stanley in 1932 (later Stanley Engineering, and then Stanley Consultants) and co-founded the furniture manufacturing company Home-O-Nize in 1944 (today the HNI Corporation), both still thriving operations to this day and based in Muscatine. And like Musser before them, the Stanleys felt compelled to pay it forward. Their establishment of the Stanley Center for Peace and Security in 1956 (founded as the Stanley Foundation) was a means for them to have an impact on critical global concerns, but also to return some good will to the city where their fortune was built.

As it concerns the new headquarters of the Stanley Center, in downtown Muscatine, we find the distinct legacies of Musser and Stanley intersecting in ways that transcend any one industry, any singular accomplishment or historical milestone. With their fortune, the Mussers built a library: a place to commune, learn, and find refuge; a place of great civic importance. The Stanleys, with their fortune, established a foundation whose mission was essentially two-fold: to help facilitate peace on the world stage, and to become a beacon of goodwill within its own community. As it concerns the latter, the Stanleys were enthusiastic supporters of the arts, and that legacy lives on — in part — through the Muscatine Art Center, which houses the three-story Stanley Gallery and happens to be attached to the original home of Laura Musser, built by her cousin and founder of the Musser Library, P.M. Musser.

Betty and Max Stanley, May 1980

This future Living Building in Muscatine represents the physical and spiritual convergence of these two legacies. "This organization has extreme financial privilege," says Catherine Elliott, a member of the Stanley Center's governance, and the great-granddaughter of Max and Betty. "From my perspective as a family member and someone who is involved in governance, we have an obligation to channel that privilege in a positive way."

MUSCATINE AND CIVIL RIGHTS

While industry and trade boomed in Muscatine in the latter half of the nineteenth century and well into the twentieth, the city would also become an unlikely yet decisive center in the struggle for African American civil rights.

Alexander Clark was born a free man in 1826, in the town of Washington, Pennsylvania, thirty miles southwest of Pittsburgh. Making his way gradually west in his teenage years, Clark was working as a bartender on a steamboat that traveled up and down the Mississippi before landing in Muscatine (then Bloomington), aged just sixteen. Clark began work as a barber, a well-esteemed profession at the time that afforded him proximity to the town's (white) industrial and business elite. In leveraging his contacts and acquired knowledge, Clark built a successful and diversified business in everything from timber sales and real estate holdings to the sale of sundry items to steamboat crews.

Clark further leveraged his social standing to become an important civic leader. Throughout the 1840s, 1850s, and 1860s, Alexander Clark's accomplishments have an impact that borders on American folklore. During the Civil War, Clark was instrumental in organizing and recruiting for Iowa's Black regiment, the 60th U.S. Colored Infantry. He was also an outspoken advocate for the desegregation of schools and Black suffrage. Clark once said, "He who has been trusted with the musket, can and ought to be trusted with the ballot." Clark was one of the founders of Muscatine's local African Methodist Episcopal (AME) church; he helped organize the local Prince Hall Mason Lodge; he was the sole agent in Iowa on behalf of Frederick Douglass' anti-slavery newspaper *The North Star* (Clark and Douglass maintained a long-standing friendship via correspondence, and they met in person at least once, at the Colored National Convention, held in Rochester, New York, in July 1853); and

Portrait of Alexander Clark, 1876

Scott House, on southwest corner of Third St. and Iowa Ave. which once housed Grammar School No. 2. (present day Welsh Apartments)

"Iowa was very much on the map for prominent Black figures of the era — Frederick Douglass, Madam C.J. Walker, James Madison Bell, and others — who travelled throughout the country and spoke out against slavery and for suffrage and for all the different rights that Black people were fighting for."

RACHELLE CHASE
Founder & Executive Director, Uniting Through History, and Author of Creating the Black Utopia of Buxton, Iowa

Unknown woman posing in front of Alexander Clark House, on its original site, circa early 20th century

he was steadily active in assisting fugitive slaves who fled Missouri and other points south via the Mississippi River.

One such case involved a young man named Jim White. White was enslaved in St. Louis, Missouri, but, with permission from his enslaver, had found work on a steamboat that traveled upriver and led him to Muscatine. Rather than return to bondage, White took refuge in Clark's home and, with his assistance, avoided capture by bounty hunters who had posed to Clark as abolitionists and claimed they would help escort White to Chicago. Clark, who was "too smart to be gulled by such an old villain," according to J.P. Walton, from his 1882 address titled "Unwritten History of Bloomington (Now Muscatine) in Early Days," helped White elude capture under cover of night. The similarities between this episode and the literary character "Jim" from *Adventures of Huckleberry Finn* have led some to speculate — with little to no evidence — that, during Mark Twain's brief time in Muscatine years later, he crossed paths with Clark (which is likely) and learned about Jim White, thus lending inspiration for the Great American Novel of the era.

In the fall of 1867, Clark tried to send his 13-year-old daughter Susan to the local secondary school. (At the time, Black children were admitted to Muscatine's segregated elementary school, but no secondary school was available to them.) When Susan was not admitted to Grammar School No. 2 — now an apartment building with ground-floor retail that stands across the street from the Stanley Center — on grounds of her race, Clark sued the Board of Directors of the Muscatine school system.

"The refusal to admit these children is a high-handed act of despotism," reads an excerpt from the September 14, 1867, edition of the *Muscatine Journal*. "It is unjustified by any principle of law or any sentiment of humanity."

Following a local court decision in Clark's favor, which was inevitably appealed by the Board of Directors, the case then came before the Iowa Supreme Court, which upheld the lower court's decision. This seminal case, *Clark v. Board of Directors*, decided in 1868, marks the first time in U.S. history that a state court declared (twice in this instance) that a school may not discriminate against students on the basis of race; it also represents the first successful school desegregation legal case in the country. According to Paul Finkelman, a legal historian, *Clark v. Board of Directors* has since been cited in forty-two separate court cases, including twice in the U.S. Supreme Court and twice in the U.S. Court of Appeals, as well as in five legal briefs filed in the far more well-known 1954 case *Brown v. Board of Education of Topeka*.

Had Clark's story ended with his business successes, or his efforts to organize Iowa's Black regiment, or his work on behalf of Black suffrage, or his singular accomplishment of helping desegregate the state's public school system, any one of these would suffice to secure his name in American lore. But his story doesn't end there. In 1879, Clark's son Alexander Jr. became the first Black graduate from the University of Iowa College of Law and the first Black person to earn a law degree west of the Mississippi. Following his son's example, five years later, Clark, at the age of 58, became the second person to earn such a degree.

Clark went on to practice law in both Iowa and Illinois. He became co-owner of *The Chicago Conservator*, the first Black-owned newspaper in Chicago. He became treasurer and chair

> *"Alexander Clark is the most important African American leader who almost no one has heard of... he is a significant figure in the history of the United States and a giant in the history of Iowa."*
>
> **PAUL FINKELMAN**
> Author, Professor, and American legal historian

Susan Clark Junior High School in Muscatine, formerly West Middle School, officially renamed in 2019

of the Executive Committee of the National Press Association, the equivalent of the present-day National Association of Black Journalists. And in 1890, after having rejected an appointment from President Ulysses S. Grant nearly ten years prior to be U.S. Ambassador to Haiti, Clark accepted an appointment from President Benjamin Harrison to be the Ambassador to Liberia. (Clark was one of four Muscatine residents to be appointed as a diplomatic envoy between 1855 — 1900.)

Like John Boepple wading into the waters of the Mississippi, Clark's venture to the West African nation was rife with hidden dangers. Succumbing to disease for which western travelers were grossly unprepared, Clark contracted fever and died less than one year into his appointment.

In Alexander Clark, we find a historical figure of enormous consequence. Adversely, we also find an individual whose distinction has been sidelined by history and whose legacy is barely recognized beyond the borders of Iowa. (In the official History of Muscatine County, published in 1911, there isn't a single mention of Clark!) And yet, Clark's example is thoroughly embedded in the DNA of American civil rights, even if it remains largely unknown to this day. The work of righting that wrong must continue, and it does through the efforts and words of Muscatine residents like Daniel G. Clark (no relation), a local historian and columnist for the *Muscatine Journal*, and Kent Sissel, a retired professor and owner/proprietor of the Alexander Clark House. That work is also reflected in the Muscatine Community School District's recent decision, in 2019, to rename the town's middle school to Susan Clark Junior High School, and in the Stanley Center's own efforts to publish two children's books about Susan Clark. The first of these, "Susie Clark: The Bravest Girl You've Ever Seen: Desegregating Iowa Schools in 1868," authored by Joshalyn Hickey-Johnson and illustrated by Hayle Calvin, both Iowa natives, was published in 2023. The second is a forthcoming historical fiction book for middle-grade students titled "The Time I Was Susan Clark," authored by Rachelle Chase.

Part of the reason Clark's example is so enduring is because of the town he made his own. Alexander Clark found great success in Muscatine, and consequently, the Stanley Center for Peace and Security now resides in what's known as the "Alexander Clark Heritage District," which was established in 2010 by the city's Historic Preservation Commission but is not recognized by Iowa or the federal government as a designated historic district.

Through "a combination of knowledge and intelligence," in the words of Paul Finkelman, Clark leveraged his success as a businessperson into further success as a civic leader, a champion of civil rights, and a purveyor of democratic ideals. In that regard, Clark's example is a forerunner of that of Max and Betty Stanley, whose own wealth, civic pride, and persistent concern for peace and justice became the bedrock on which the Stanley Center was built. As Max noted in an address on global citizenship that he delivered at the University of Dubuque, in 1983, "Wisdom and intelligence, cooperation and coordination, innovation and determination are all necessary, but they are not enough. We need greater compassion. Understanding, respect, and love are needed to accommodate our differences and unite our efforts to enhance the livability and grandeur of this tiny ball spinning in space."

1901

1960s

THIRD AND IOWA

As typical urban plots go, the site for the Stanley Center's Living Building boasts few qualities of note to the naked eye. However, what transpired on this corner — structurally, ecologically, and historically — makes the corner of Third Street and Iowa Avenue one of the most important intersections in Muscatine's history.

As far back as 1877, a small workshop stood on the site, operated by a carpenter named Henry Madden, who is listed in the city directory as the occupant of Third and Iowa. Such ancillary businesses were typical of the time, popping up around town to service the area's booming lumber industry. Meanwhile, on the western end of the same city block, at the corner of Third and Chestnut Streets, stood a handsome two-story brick house built in the customary Federal Style, the property of one Alexander Clark. (The brick structure replaced Clark's previous home, a wood-framed house that occupied the same footprint, which had been destroyed in a fire, likely caused by arson.) Only a few years later, Third and Iowa's footprint became denser, with a livery and feed stables, small warehouses, a blacksmith shop, and a carpentry shop all occupying the site. Adding to the tenuous and haphazard development patterns of the site, cutting diagonally through the plot — flowing approximately west to east — was a circuitous creek bed that sat dry in the summers.

Small business operations came and went, each a remnant of an era of great industrial growth, but otherwise held in check by ecological conditions that precluded the building of something grand and with real civic presence. By 1892, the creek bed had been patched over and diverted into a subterranean culvert. Three years later, a fire destroyed much of the site, save for the two-story building that housed the blacksmith shop on the ground floor and the carpenter's workshop above it, which would remain in place until the turn of the century.

Musser Library through the years: the Patton, Fisher & Miller-designed library, completed in 1901; the original structure along with the 1965 extension; by 1971, the original 1901 building had been razed and replaced with the structure that stands today

In 1897, a small library association was formed in Muscatine, comprising a grouping of civic and business leaders that included P.M. Musser. At the same time, the city's public book collection was warehoused in the basement of the local high school, and the association was intent on remedying that with a new building. The only issue was where the money for such an endeavor might come from.

The next year, J.P. Walton, a local businessman, carpenter, and president of the Muscatine County Old Settlers' Society, published a newspaper article in which he wrote, "We have a suggestion to make to the Mussers. The city needs a public building for an academy of science and a library. If they would now in [their] lifetime devote a few thousand dollars and put up a Musser Building to be used for such purposes, it would place the Musser name on perpetual record." This public plea evidently had the desired effect. By 1900, a newly proposed library was announced in the Christmas edition of Muscatine's The Saturday Mail, with a rendered design by Chicago architects Patton, Fisher & Miller (the first of many libraries for the firm), and an initial $30,000 donation from Musser, who by then owned the site and soon after transferred ownership of the land and building to the city. Ground was broken the following year and construction commenced at the corner of Third and Iowa.

The P.M. Musser Library opened to the public on February 19, 1902. With a façade of red sandstone, a red slate roof, steam heat, and electric lighting (Muscatine's power plant opened the previous year), the new three-story building was undeniably stately but without the flash embodied in other stone structures that went up during America's Gilded Age. Its deep setback and dramatic curb appeal communicated its purpose as a place of learning. "It was a beautiful old building, and you would walk up these big steps to the front door," Muscatine resident and local archivist Sheila Chaudoin, a long-time library employee, commented in a 2016 article in the Muscatine Journal. "And on the fourth or fifth step up was the date of the building, 1901. As kids we called that the 'poison step,' so you had to step over it to get into the building."

As beloved as this library was, it also came with issues. At first, the central concern was capacity. An extension was added to the

Left: Mark Seaman and Robert Fiedler investigating the contents of the 1971 time capsule

Right: The expanded time capsule being returned to 304 Iowa Avenue, early 2023

back of the building in 1965, with funds donated by P.M.'s son, C.R. Musser. But in 1970, an engineer's report indicated that the original 1901 building was structurally compromised. In other words, it was slowly sinking. The building that replaced the original library, along with the 1965 addition, are what today comprise the bones of the Stanley Center's new home.

Unlike its predecessor, the 1971 Musser Public Library was not so aesthetically beloved. The building boasted a ground floor atrium that received little daylight, odd-shaped balconies that few patrons ever used, and the curb appeal of your average parking garage. It was brutal without the benefit of being Brutalist. Thankfully, the institution took some care to honor its history. Prior to razing the old building, the library's director, James White, led an effort to salvage the 1901 "poison step" and make it the cornerstone for the new building. An airtight metal time capsule was placed inside the cornerstone, containing, among other items, a letter written by White, addressed "To Whom It May Concern": "The building was [sic] are passing on to you was planned to the best of our abilities. Only you, in your time, can judge whether we planned carefully. We intended to build for twenty years of expansion. Time will tell whether we were correct or not."

Time proved him correct, plus another twenty-five years. By 2017, space constraints and "multiple building issues," according to the library's website, forced the library to explore options for a new home, which they found in the form of the recently vacated headquarters of furniture manufacturer HNI Corporation, which donated their old building and moved across the street to new digs on East Second Street, in a renovated warehouse designed by Neumann Monson Architects. (HNI, a company founded by Max Stanley, also played a key role in furnishing the Neumann Monson-designed Stanley Center.)

Third and Iowa languished. This odd building, with its bric-a-brac triangles jutting out from the second floor and a (presumably) forgotten cornerstone, had become a puzzle for the city to solve. It may have been an eyesore by contemporary standards, according to some, but there was no denying the structural superiority of the 1971 building over its more attractive turn-of-the-century predecessor. So, tear it down? But replace it with what? Renovate it? But for what use? Either way, no one had a good answer.

Then sometime in 2019, the board of directors of the Stanley Center, as well as Center leadership, began a process of self-examination. Since Max Stanley's passing in 1984, the organization had undergone several changes, forged new partnerships, and launched new initiatives. Its internal ranks grew as well, and it formally adopted climate change as a core focus of its work. With new growth inevitably comes growing pains, and as the second decade of the twenty-first century approached its end, two things had become apparent: Stanley Center staff and leadership knew their organization needed a new home, and the board of directors (which maintains strong representation from Stanley family members and descendants) expressed their wishes that the foundation remain in Muscatine and become a place where people would want to come to work every day.

The former Musser Library building at Third and Iowa wasn't the obvious choice, and by many accounts it wasn't the easy choice. But it was undeniably the right choice.

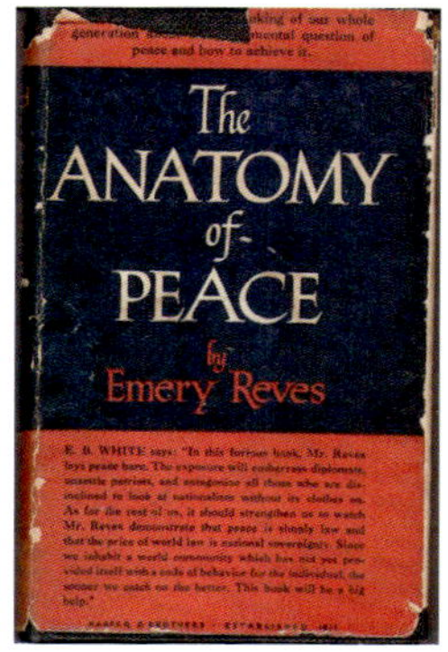

"The Anatomy of Peace," first published in 1945, and portrait of its author Emery Reves

THE ANATOMY OF PEACE

It all started with a book. In 1947, Max and Betty's eldest child Dave was attending the University of Iowa as a freshman. While at school, Dave became enamored with the work of the United World Federalists (UWF), a U.S.-based association that sought global peace through "enforceable world law," achieved principally by emboldening the United Nations to have greater oversight in international affairs.

Dave brought home a copy of *Anatomy of Peace*, written by Emery Reves, and implored his dad to read it. Reves' book was a pivotal work of political and social theory that embodied world federalist sentiment in the wake of World War II. It also demonstrated astonishing optimism in the face of human civilization. In one passage that surely resonated with Max, Reves wrote, "For some thousands of years we have been struggling for peace. That we have never reached our goal does not prove that peace is unattainable. But it does prove that the means and methods by which we have tried to achieve it are inadequate."

This was a turning point for Max Stanley. That same year he too became involved in the UWF and, according to his youngest son Dick (who succeeded his father as the Center's president in 1984), a "fervent believer" in its mission to strengthen international law and prevent another world war. Typical for someone whose probity and enterprising ways helped make his dual business ventures Stanley Engineering (later Stanley Consultants) and Home-O-Nize (later HNI Corporation) such successes, Max's ascent and growing influence within the UWF was unsurprising. He served non-consecutive terms as UWF's president, from 1954-56, and again from 1964-66; he oversaw the organization's growth to include 700 chapters and a membership of 50,000 strong; he testified before a Senate subcommittee to discuss proposed revisions to the UN Charter; he was invited to the White House.

This newfound purpose also translated into the dealings of his engineering firm. In 1956, Stanley Consultants began to operate overseas, in Liberia. (The shared connection to Liberia between Max Stanley and Alexander Clark is happenstance.) The firm was initially retained to design a diesel power plant, which subsequently led to other contracts to update a commercial shipping harbor and supervise the construction of a $25 million hydroelectric plant, among other jobs. According to Max's biographer Ros Jensen, the new international division launched in West Africa was "partly an outgrowth of the world outlook that deepened after he joined UWF." While Max was "content to keep Muscatine the center of his universe ... he was impelled to reach beyond the provincial boundaries to satisfy his interests in the world's well-being." For his service, Max Stanley would eventually receive the Order of African Redemption from the government of Liberia.

That same year, Max published his book, *Waging Peace: A Businessman Looks at United States Foreign Policy*, in which he adopts the case for the U.S. maintaining a strong military presence as a deterrent to global war — a position that he softened on considerably in his later years.

It didn't take long for Max to grow disheartened by what he considered the UN's inability to influence international policy and, in his son Dick's words, "excessively doctrinaire" positions of the UWF. Together, Max and Betty sought a different tack, one focused on grassroots work, grounded in compassion, and tethered to the philosophy that people generally wish to do good for the world, often despite their perceived diplomatic obligations.

The Stanley Center for Peace and Security (formerly the Stanley Foundation, which changed its name in 2019) was established in December 1956, with $250 in cash and 160 shares of Home-O-Nize common stock, valued at about $10 per share. (While Stanley Consultants established Max's reputation as an engineer and public citizen of the highest order, the success of Home-O-Nize made the Stanleys millionaires several times over. Thanks to the furniture manufacturer's continued growth and rising stock price, the Center's net worth was nearly $20 million the year Max died.) All three of the Stanley children, Dave, Dick, and Jane, sat on the founding board of directors, alongside their parents. In its first years, the Center's activities were largely philanthropic. According to Ros Jensen, the organization "concentrated mainly on giving scholarships to high school and college students and donating funds to a number of charitable, educational, and religious organizations with international programs." But such activities seemed designed to keep the Stanley Center at arm's length, serving more as a conduit for good rather than its prime instrument. Obviously, this was neither Max's nor Betty's design.

In later years, the Stanleys would hone their organization's focus in several arenas, both international and hyperlocal. Max wanted to bring people from all corners of the globe together, not as self-interested emissaries brimming with nationalist fervor, but as human beings wholly committed to "securing peace with freedom and justice," a term he adopted as his lifelong mantra after reading Reves' book. Betty's passion was education, and through the Center she sought to bring the world to Muscatine and its schoolchildren with a program known as Project Enrichment (discussed further in the next chapter).

In his book *Waging Peace*, Max took a somewhat hawkish position when it came to the Soviet Union, pitting back-to-back chapters titled "The Communist Plan" and "Our Plans." "Under this dictatorship, both civil and international war will be used to establish power ... [Communism] uses war and conflict to expand itself," he wrote. But in later years, owing to the important work of the Center and its philosophy of collaboration, Max's world outlook became more empathetic. Max and Betty traveled to all corners of the globe several times over — particularly throughout Africa and Central Asia — and had occasion to break bread with people from all faiths and political persuasions. He bristled at Cold War-related posturing from both world superpowers and the kinds of diplomatic squabbles he deemed more petty than strategic. He came to see peace not as something to be "waged" (implying its achievement by means of treaties or ceasefires) but managed and tended to, through partnerships and mutual understanding.

In yet another passage from *Anatomy of Peace* that surely resonated with Max, Reves wrote, "All the planners of lasting peace believe that theirs is the magic formula; that they can make something work which never has worked; that after the failure of thousands of peace treaties they can draft one that will prevent war."

Max wrote another book, *Managing Global Problems*, which was published in 1979. In its pages we find the words of someone who has evolved from a businessman considering foreign policy (an outsider) to a seasoned, learned, and well-versed ambassador for peace (the insider). His biographer wrote that, "Beginning in the seventies, [Max] put increasing emphasis on global problems that no longer were peripheral to peace and disarmament." In the same breath when expressing concern over nuclear war, Max also spoke "on the dangers of various forms of pollution, unchecked population growth, diminishing natural resources, and the serious imbalance of financial resources between the haves and have-nots."

In Max's own words, "protecting and managing the biosphere" became one of his singular preoccupations, especially in his later years. Max Stanley was known as someone who did his homework and was keen on exploring solutions, not just pontificating on problems. In *Managing Global Problems*, he foresaw societies and a world in dire need of greater conservation and renewable energy. "Water resources deserve attention to assure efficient development and reuse," he wrote. "To meet expanding energy requirements, we must tap the inexhaustible energy of the sun, the heat within the earth's core, the winds, and other natural forces."

In another passage he wrote, "It takes wisdom, courage, and strength to recognize the inevitability of change, to adjust to it, and to grasp opportunities. A keen sense of timing and willingness to innovate are needed."

In these sage words we find the core principles of the Stanley Center for Peace and Security, past and present.

Top: Max in his office, circa 1980 *Middle:* Virginia and Ambassador Carlos Romulo of the Philippines with Betty and Max Stanley at the World Congress of World Federalists, San Francisco Hilton, June 1965 *Bottom:* Stanley Center staff at work, circa mid-1980s, with Dick Stanley at the head of the table

"How aspirational and pie-in-the-sky was this vision, that an organization based in Iowa and devoted to influencing global governance, could have such an impact? That was a bold move in 1956, and this Living Building is a bold move in 2023."

CATHERINE ELLIOTT
Member of Governance, Stanley Center for Peace and Security

THE CENTER EVOLVES

Among Max Stanley's numerous dedications and passions, both global and local, one included his long-time involvement with the Boy Scouts of America.

For many years he led the troop to which both his sons belonged. His eldest Dave once recalled a time when the troop was picking up their campsite and the scoutmaster instructed the boys to clean up litter previously left by other campers. According to Dave's account, after some grumbling from the troops, his father remarked, "Wherever you go, you leave it better than it was when you found it."

Entering the 1960s, a decade largely defined by international armed conflicts, civil unrest, and rising Cold War tensions, the Stanley Center's mission and guiding principles would be put to the test. In the face of such upheaval, Max and Betty always resolved to meet challenges head on, through a multilateralist approach with governmental and non-governmental actors alike. Consequently, Max gradually stepped away from his daily responsibilities at Stanley Consultants and HNI to focus an ever-increasing share of his time on the Center's expanding efforts.

In 1960, the Stanley Center held its inaugural Strategy for Peace Conference (SPC) at Arden House, in Harriman, New York. The annual conference — held at Airlie House in Warrenton, Virginia, starting in its second year — has historically brought together experts from the public and private sectors to discuss global challenges and policy solutions. The outcomes of these conferences contributed in some measure, either directly or indirectly, to the formation of the UN's Environment Program (UNEP), the U.S. Arms Control and Disarmament Agency (dissolved in 1999), and international negotiations over the peaceful exploration of outer space, among other developments. Max was adamant that these gatherings at Airlie House — where it was not uncommon for representatives from the World Bank, the International Monetary Fund, and other international organizations and national governments to be in attendance — not merely be shallow demonstrations with no follow through. He wanted the conversations and connections made at the SPC to be the catalyst for nuclear nonproliferation and global negotiations that would bring some stability to international economic and energy policy. He cared for substance and real change, not paying lip service to the status quo.

In keeping with Max's world federalist leanings, the Stanley Center also established the United Nations of the Next Decade, in 1965, an annual week-long conference that brought together government officials and academics from around the globe to address how to make the UN more effective on the world stage. (The conference was discontinued in 2008.) And in 1970, the Center established the annual UN Global Issues Conference (founded as the UN Procedures Conference and initially held at Arden House before moving to Tarrytown, New York), which convened international researchers and subject matter experts to address a range of global concerns, from genocide prevention to climate change.

The Center became wholly committed to publicly promoting a free and independent press by sponsoring a series of annual programs, workshops, and fellowships around the world that ensure journalists have access to information and training, a practice that continues today. Tangentially, the Stanley Center entered the arena of mass media production. In 1975, it became the publisher of *World Press Review*, a monthly magazine that showcased articles and analyses on global issues that were originally published outside the United States. In 1980, the Center began producing the nationally syndicated *Common Ground*, a weekly public radio program

"Max Stanley very much believed that educated citizens are important to our process and to what the Stanley Center sets out to accomplish."

JENNIFER SMYSER
Vice President and Director of Policy Programming Strategy, Stanley Center for Peace and Security

devoted to in-depth coverage of critical international concerns, including growing concerns over environmental issues. (The Stanley Center's current CEO Keith Porter previously worked as a producer and co-host for the program.)

Well into the Stanley Center's third decade of existence, Max and Betty were bona fide internationalists, with a collective influence that extended well beyond the borders of Iowa and the nation, to paraphrase Max's 1984 obituary in the *Muscatine Journal.* At several points, Max was encouraged by outside parties — to no avail, thankfully — to base his business interests and foundation activities in larger markets to capitalize on that influence; Stanley Consultants to either Chicago or St. Louis, the Stanley Center to either Washington, D.C., or New York. The couple had also built, over the course of the 1970s and early 1980s, one of the world's most highly regarded collections of African art, numbering approximately 380 pieces. What began as a singular passion of Betty's during the couple's early travels to Liberia quickly turned into a mutual fascination and transformed the Stanleys into sought-after collectors by the art world elite. This new venture, however, was not the pursuit of dilettantes. The bulk of their collection is composed of wooden masks, sculptural figures, and other tools, mostly religious or ceremonial in nature, stemming from all corners of the continent; it is the work of craftspeople working (mostly) in agrarian settings, designing objects with utility and purpose. As their knowledge and collection grew, it was of critical importance to Max and Betty that each piece and artifact reflect cultural beliefs, functions, and routines of daily life. (The Stanley Collection of African Art is now housed at the University of Iowa's Stanley Museum of Art.)

Max and Betty established a second foundation in 1979, called E&M Charities (for Elizabeth and Maxwell), which enabled them to award grants to various religious, educational, and charitable organizations — a practice the Stanley Center does not engage

Top: Selected pieces from Max and Betty's collection *Middle:* Max and Betty's living room in Muscatine *Bottom:* Exhibit at the Stanley Gallery, Muscatine Art Center, circa 1980

"A colleague of mine once posed a question to Max, something to the effect of, 'You're seen as a man who's rich and powerful and well connected. But what can ordinary people do to make a difference in the world?' Max then looks at him and replies, 'Well, one person can't do much. But one plus one plus one adds up."

DANIEL G. CLARK
Historian, Journalist, and former employee of the Stanley Center for Peace and Security

in. But through it all, the couple maintained a fierce loyalty to their homebase of Muscatine, something they preferred to demonstrate with action and not idle talk. In this era, the Stanley Center's commitment to its home city was evidenced by Betty's Project Enrichment initiative. Launched in 1971, the program's chief goal was to nurture good citizenship and global awareness through education. Local students gained access not only to books, films, and other media focused on global issues, but to authors, civic leaders, and international dignitaries who visited Muscatine schools and met with students face to face. Today, Betty's program lives on through the Stanley Center's Global Education program. Among its numerous efforts, which include enriching the Model UN at Muscatine High School, the Stanley Center also provides the Catherine Miller Explorer Awards to Muscatine teachers, an international summer study tour opportunity that is designed to, according to the Center's website, enhance "their ability to support students from diverse backgrounds, add new perspective to their classrooms, and help them become stronger educators."

The Stanley Center for Peace and Security has never wavered in its faith in the promise of effective global governance. If anything, the Center's evolution has been continual, seeking new avenues (and new collaborators) to help solve what ails the world. It has also stayed true to the principles of its founders Max and Betty Stanley. True partners in both life and philanthropy, the Stanleys believed that impactful change can start with grassroots efforts, which also gives credence to their preference to remain in Muscatine and not decamp to the shores of the Potomac in Washington, D.C., or to Midtown Manhattan, in the shadow of UN Headquarters. Max and Betty were citizens of the world, yet they never stopped enriching their own community. They were the embodiment of thinking globally and acting locally. This example lives on in the Stanley Center and its activities today. And it comes to life in new ways through the Center's efforts to create a fully regenerative and restorative Living Building in Downtown Muscatine.

Top: Stanley Center's Krista Regennitter addressing students *Middle:* Gathering at Stanley Center's 60th Strategy for Peace Conference, Warrenton, VA, October 17, 2019 *Bottom:* Daniel Cuevas of the Jesuit Refugee Service (left) and Keith Porter observe the Colombian border with Venezuela near Cúcuta, Colombia, April 26, 2019

Participants gather for a panel discussion during War Stories Peace Stories: Peace, Conflict, and the Media, co-organized by the Stanley Center and The Times Center, New York, New York, April 11, 2018.

A NEW ERA FOR THE STANLEY CENTER

The Stanley Center's CEO Keith Porter, a veteran of the organization who has been at the helm since 2013, recalls the moment when the issue of climate change was formally adopted as part of the organization's strategic plan and scope of action.

Porter and some of his Stanley Center colleagues attended the UN Climate Change Conference (COP21) in Paris, in 2015, where the pivotal Paris Agreement was formulated (and became a legally binding international treaty the following year). The most well-known goal of the Agreement committed nations to "holding the increase in the global average temperature to well below 2°C above pre-industrial levels and pursuing efforts to limit the temperature increase to 1.5°C above pre-industrial levels." (As of 2023, the warmest year on record, global temperatures have already increased to 1.2 degrees above pre-industrial levels.)

Porter recalls, "There were several countries that pushed very hard for the limit of 1.5. We felt it would take a lot of effort to keep that limit in front of policymakers and the media." The Stanley Center was joined by other organizations following COP21 that set their sights on keeping the civic discourse focused on 1.5. "We said, 'We're going to hold people's feet to the fire on this. We're going to talk to policymakers. We're going to make policy recommendations in line with the most ambitious part of the Paris Agreement.' And from there, we developed our first work plan to develop and promote transformational pathways to a 1.5 limit."

One big piece to the emissions puzzle is reducing the world's overall reliance on coal. (China remains the world leader in GHG emissions from coal combustion, with India and the United States coming in a distant second and third, respectively.) The Stanley Center endeavored, initially, to focus on this chief source of emissions. And in keeping

"There are several design features of this building that are driven by function. That symbiosis drives a lot of questions about what makes a Living Building and how it becomes a reality. And hopefully those questions lead to learning for all of us."

KEITH PORTER
CEO and President, Stanley Center for Peace and Security

"It was an educational process. As we went through the analytical and planning steps to realizing a more sustainable building, we came to the collective realization that we shouldn't cut ourselves short. We said to ourselves, 'Let's go for the maximum; let's see what can truly be achieved!"

WARREN HEIDBREDER
Member of Governance, Stanley Center for Peace and Security

with Max Stanley's pragmatism and business mind, they framed the coal issue as a question of investment. "This wasn't about pressuring governments individually to be good stewards," Porter says. "This was about financing and liability. Anyone investing in coal today will have stranded assets on their books within ten years. That's bad business."

In the months and years following the Paris Agreement, the challenges to combating climate change began to come into sharper focus for a greater portion of the public, particularly for laypeople and professionals not working in sectors tied to climate science or the building and energy sectors. Terms like "carbon footprint" and "net-zero" entered the lexicon, and the differences between operational versus embodied carbon emissions were laid bare for anyone showing an interest. Consequently, the Stanley Center developed a work plan focused on three key strategies: 'Examining and accelerating promising and lesser-explored pathways', 'Create opportunities out of international political moments', and 'Looking for solutions outside of climate policy.'

That last one stands out. "You can either fix problems, or you can transform minds," Porter says, drawing a distinction between applying band-aids and instituting systemic change. (Porter maintains that both have their place.) Despite best intentions, achieving the latter rarely is the result of meaningful policy change. Fixing problems, particularly as it concerns climate change, often requires new laws, treaties, and memorandums; it requires cap-and-trade agreements, enforcing building codes and performance standards, and passing municipal ordinances. Globally, fixing these problems requires bringing together thousands of international delegates from more than one hundred countries, every year, and signing yet another treaty that will be broken by a large fraction of those in attendance. Transforming minds, on the other hand, tends to be a local matter.

"Everything that's global is local somewhere," says Catherine Elliott, a member of the Stanley Center's governance, echoing one of her great-grandparents' (Max and Betty) core tenets and, in many respects, the Stanley Center's founding charter. Elliott believes that meaningful climate action is a matter of "boiling down." As she describes it, governments get involved, followed by NGOs, then staff working in other organizations who advocate for new resources, which find their way into cities and towns and eventually into schools, where teachers and students gain access to new sets of knowledge. "It all comes down to a very local point," Elliott says.

Undoubtedly, making transformative change happen is integral to the Stanley Center's broader mission. The completion of the organization's future Living Building presents a tremendous opportunity to lead by example, both on the world stage and right in their hometown. "Part of the Living Building Challenge is considering your impact on your neighbors," says Jennifer Smyser, Vice President and Director of Policy Programming Strategy at the Stanley Center. She also concedes, in classic Midwest pragmatism, that "we don't need people to know what the Stanley Center is to do our work ... we're not looking to build a huge public audience." And yet, many of the people who can (and will) benefit from this Living Building — and of the Center's work to realize a peaceful, equitable, and livable climate future — are those who call Muscatine home. Not policy makers or UN delegates, but the school children and private citizens of this small industrial town on the banks of the Mississippi River.

(L-R) Stanley Center Board Vice-Chair Brian Hanson, UN Secretary-General Ban Ki-moon, and Stanley Center CEO Keith Porter, January 14, 2013

"This is our chance to do some of that transformative change, to have a building that not only generates enough energy for its *own* use, but gives back to our community," Smyser says. What is ultimately paid forward will be both tangible, in the form of food, water, and energy, and intangible, in the form of knowledge, education, and the experience that comes with operating a restorative and highly visible building in the center of town.

To Smyser's earlier point, the Stanley Center doesn't need a billboard. And while various media have been leveraged in the past to promote the organization's work and message, namely the radio program *Common Ground* (broadcast from 1980 to 2004), they have kept a low profile, just as the Stanleys did. This is about to change.

For the first time in the organization's history, it will have a home with real presence and street-level visibility. At the intersection of Third and Iowa, the building's overhang of solar panels proclaims this otherwise modest structure as a source of renewable energy and high-performance design; two giant cisterns sited on the corner where the Center meets the alleyway communicate good stewardship. From its cladding and glazing systems to its courtyard gardens bursting with edible vegetation, this once abandoned public library building has been redesigned with pride and boldness. Its very appearance projects light, growth, and achievement. Going forward, the Stanley Center's new home is sure to attract attention.

"The Stanley Center is now visible in a way that it has never been. It has always been sort of hidden in an upstairs location, away from the public. This building gives us so many opportunities to be a more effective member of the community."

LYNNE E. STANLEY
Chair, Board of Directors, Stanley Center for Peace and Security

THIRD AND IOWA, REVISITED

Making the Stanley Center's new home on the corner of Third and Iowa was not the obvious choice. The former library building was bulky, its exterior was drab, and the interior did not receive a lot of natural light. Further, the nearly 24,000-square-foot building provided the Stanley Center and its staff of twenty-five with far more workspace than they needed or were accustomed to.

Prior to 2020 and the onset of the Covid-19 pandemic, the Stanley Center operated out of the top two floors of the Laurel Building, a six-story office tower built in the early twentieth century and designed by Chicago architect Paul V. Hyland, located just down the block at Second Street and Iowa Avenue. Stanley Center staff have described their former digs at the Laurel as "disconnected" and "darkened," with cubicle farms and conference rooms under fluorescent lights, but "the river views were excellent."

"We launched our new brand in November 2019, and four months later the pandemic hit," says Mark Seaman. "In those four months, we painted the walls in our new colors, we put up our new logo, our brand manifesto, enhanced our conference areas ... and then we didn't go into the office." Anyone with not-so-distant memories of the corporate workplace life can likely empathize with this plight.

"It was a yearlong process ... trying to address how our communications could be more effective," Seaman told the *Muscatine Journal* in October 2019, on the cusp of launching their new brand and transitioning from the Stanley Foundation to the Stanley Center for Peace and Security. Key to gaining that clarity was the organization's need to differentiate itself from the more well-known Stanley Consultants. (Bizarrely, this confusion was made plain in the same 2019 article in which Seaman was quoted, when a photo of Stanley Consultants' office exterior, which sits on the opposite corner of Third and Iowa, was mistakenly attributed to the 'Stanley Foundation.')

Of course, the Stanley Center's rebrand involved much more than a new name and color scheme. It was also an "evolution of our governance," according to Keith Porter. It was an opportunity to revisit strategic goals and what it meant to operate this organization and maintain its desired level of global influence, all while staying put in this river town of 20,000 people. That balancing act was "challenging," says Porter. The Stanley Center's governing leadership acknowledged those challenges while making clear their wishes that the organization remain in Iowa.

As the thinking went, uprooting operations and moving to a more metropolitan setting would make all the sense in the world on paper. It would also largely undo the legacy and example established by Max and Betty Stanley, whose collective hearts never strayed far from Muscatine.

In all fairness, hybrid work had been part and parcel of the Stanley Center's work culture for many years, and the pre-Covid era was no exception. People might come into the office two or three days a week, recalls Jennifer Smyser. "We knew we could do the work from anywhere," she says, because many staff's duties required extensive travel. So, while Covid may have put an immediate halt to all things travel related, the resulting global shift to working from home, at least for office workers, did not severely impact Stanley Center's ability to operate, only how to operate.

The only rub was, in tandem with the Stanley Center's rebranding effort, they were also actively looking for a new home. They were starting to meet with architecture firms and tour potential sites around town. Both vacant lots and existing buildings were on the table. It was settled; their current lease would not be renewed. The fifth and sixth floors of the Laurel Building may have offered panoramic views of the river and those Muscatine sunsets immortalized in the pages of Mark Twain's *Life on the Mississippi*, but it fell well short of providing the organization with a supportive and collaborative space worthy of its work.

"As we went about planning to move, the way we want to work was converging with where we wanted to work. And that process manifested a very different kind of workspace for us," Smyser says.

Pandemic or not, one truth was laid bare: the Stanley Center of the 2020s is a decidedly different kind of organization than the one Max and Betty built in the 1950s and 1960s. Which is not to say the organization had undergone a philosophical shift. Quite the opposite. The Stanley Center for Peace and Security has remained steadfast in heeding the words of Emery Reves, who wrote in *Anatomy of Peace*, "There is not the slightest hope that we can possibly solve any of the vital problems of our generation until we rise above dogmatic nation-centric conceptions and realize that, in order to understand the political, economic and social problems of this highly integrated and industrialized world, we have to shift our standpoint and see all the nations and national matters in motion, in their interrelated functions, rotating according to the same laws without any fixed points created by our own imagination for our own convenience."

Likewise, the organization continues to practice the highest ideals of international diplomacy in service of peace and security for all. This is evidenced by the Stanley Center integrating climate action into its strategic plan in the days and months following the 2015 UN Climate Change Conference (COP21) and the signing of the historic Paris Agreement.

Interior of Musser Library, circa 2010

While the Stanley Center and its staff may still embody their founders' modesty and preference to avoid the limelight, waging a decisive effort to combat global warming does require, for better or worse, greater visibility. It also demands accountability from those who choose to preach the gospel of climate action. An organization that publicly aims to realize "a peaceful, equitable, and livable climate future," but doesn't practice those same values in its place of work, undermines its own mission. Therefore, it was important that the Stanley Center's new home be a place that embodied the holistic benchmarks that comprise the Living Building Challenge. It needed to be a place of efficiency and equity and regeneration, a place that supported health and happiness and a spirit of collaboration. It needed to be of its community, front and center.

On its face, a clunky old library building doesn't necessarily scream 'opportunity!' And many of the pieces to achieving what Keith Porter calls a "values-forward" building that "matches the way we work and the way we want to work" weren't exactly hiding in plain sight. It's true that Third and Iowa wasn't an easy sell, but the selection of this site for the Center's headquarters was a matter of upholding the organization's values and giving back; it was an opportunity to invest in the community. Practically speaking, adapting the former Musser Library to become a Living Building was an enormous challenge. But overcoming that challenge and manifesting the organization's values in built form were well worth the effort.

PART II

Finding A New Home

"If we're going to build or renovate a building, we're going to do it in a way that matches what we preach to the world."

KEITH PORTER
CEO and President, Stanley Center for Peace and Security

GETTING STARTED

As beautiful and charming as Muscatine, Iowa is, the city remains an unlikely place to find the headquarters of a globally minded organization focused on international policy and addressing significant global challenges.

Still, just as the Stanleys had become pillars of their community, so too had Muscatine become part of their namesake organization's DNA. Stanley Center CEO Keith Porter doesn't shy away from pointing this out.

"Our board has made a real commitment to Iowa as the location for our organization," he says. "We talked a lot over the years about the pros and cons of how our location matches up with our mission. And the fact is, the Stanley Center exists from an endowment that was generated from local wealth, and by the work and the people in this area. All things considered, remaining in — and deepening our commitment to — Muscatine is the right decision."

But that decision does come with its own challenges, acknowledges Porter, not the least of which involves being based in an off-the-beaten-path city while occupying an inconspicuously located office space. And since one of those factors was fixed, something else had to give.

The top two floors of the Laurel Building were always conducive to, according to Porter and Mark Seaman, the kind of "heads down" and "nose to the grindstone" work that is often required of Stanley Center staff. At the same time, the organization had undergone a comprehensive review of their culture and values. They had intensified their commitment to collaboration, both externally and internally. The Stanley Center had also established a significant and very public resolution to make mitigating climate change one of its core issues. Forging ahead while staying cloistered and heads down in the top floors of an office building was incompatible.

Porter cites the need for a more "intentional" kind of workplace, a space "designed for people to come together and feel comfortable and welcome. That really got us thinking about our new headquarters."

In the beginning, the possibility of leasing space for the Stanley Center's new headquarters seemed feasible, if not probable. In the collective minds of the organization's employees, it was all just a matter of making the necessary upgrades that evoked that sense of purpose. But a final decision on the type of space to pursue remained far off. New construction and retrofits of existing buildings in town were being considered as well.

The Stanley Center team started by dipping their toes in the water, so to speak. An internal staff committee, dubbed the Blue Skies group, was asked to "dream big" as they developed a wish list for new office space. The resulting Blue Skies Memo, released in November 2019, recommended that the Stanley Center needed both private offices and ample space for collaborative discussions, an inviting communal kitchen and eating area, outdoor meeting space, lots of plants and natural light, and spaces to promote fitness and overall well-being. Further, the memo emphasized the need to promote environmental stewardship, energy efficiency, recycling, and composting.

Stanley Center leadership reached out to a few architecture firms to get a sense of who was best equipped to help their organization achieve these goals and lend design vision to the organization's wants and needs. Porter and his team knew how they worked in the past and how they wanted to work going forward. But the larger considerations, according to Porter, were

Banquette seating in the kitchen area helps conceal the raised culvert, atop the original creek bed "what would be best for the community, and what would be most in line with our values?"

Therein lies the heart of the matter. The specifics of coming to work in a sustainably designed, high-performance building had yet to be addressed; nor had the topics of air quality, biophilia, water conservation, operational emissions, embodied carbon, Red List materials, or anything remotely related to regenerative design been considered. All of that would come later. Which isn't to say such considerations were moot. Ever since COP21 and the passage of the Paris Agreement, the Stanley Center team had been advocating strongly to limit the global temperature increase to 1.5 degrees Celsius.

"If we're going to build or renovate a building, we're going to do it in a way that matches what we preach to the world," Porter says. For the time being, they were focused on finding solutions that aligned with their organization's values, and the right design partner to come up with those solutions.

"Our core values are things that we use all the time when making decisions. They're not just platitudes we put on the wall. These are things that we consistently revisit and revise. And when Neumann Monson Architects came to us with a values-based presentation, showing how their core values intersected with ours, that was deeply meaningful to us."

"We talked a lot over the years about the pros and cons of how our location matches up with our mission. And the fact is, the Stanley Center exists from an endowment that was generated from local wealth, and by the work and the people in this area. All things considered, remaining in — and deepening our commitment to — Muscatine is the right decision."

KEITH PORTER
CEO and President, Stanley Center for Peace and Security

Stanley Center staff in their office in the Laurel Building, which they occupied up until early 2020

SHARED VALUES

Left: Visioning workshop held at HNI, March 2020
Right: Site tour of the vacant Musser Library

Was it love at first sight? Perhaps. But one thing is for certain. In forging the partnership that would eventually produce the Stanley Center's future Living Building, and one of the most high-performing buildings in the entire American Midwest, it only took one meeting. Sort of.

That meeting occurred on December 19, 2019. As Porter recalls, "Neumann Monson came to us and said, 'We get you. We share a lot of your values. And we think we can help you create a space that is a physical manifestation of those values.'"

Neumann Monson had no sense up front of what the project budget was, but they knew their potential client had high aspirations and were wholly committed to creating a building, in one form or another, that gave back to Muscatine and contributed to people's health and happiness. What the Stanley Center needed, to Neumann Monson President Tim Schroeder's thinking, was the right design partner who could shed light on how to accomplish that, but without applying overt pressure. Their presentation outlined standards and guidelines for several sustainable building certification programs, including the Living Building Challenge. "We addressed all the potential [hard and soft] costs associated with the project, and how they can escalate, depending on what they wanted to achieve," recalls Schroeder, himself a native of Muscatine. "That was the beginning of the conversation" that led to the Stanley Center pursuing a Living Building.

Once the contract was signed between respective parties, the Stanley Center formed, at Neumann Monson's suggestion, an Office Space Steering Committee, which became the beating heart of the decision-making process. Also at the firm's suggestion, seven focus teams were formed to tackle many of the topics addressed in the Living Building Challenge's Petals and Imperatives. The focus teams, each of which included a member of the steering committee, were Public Education, Culture, Transportation, Facilities & Operations, Urban Agriculture, Technology, and Relocation. According to Neumann Monson Principal Sally Obernolte, the formation of these groups was "critical for creating

"Neumann Monson came to us and said, 'We get you. We share a lot of your values. And we think we can help you create a space that is a physical manifestation of those values.'"

KEITH PORTER
CEO and President, Stanley Center for Peace and Security

"We addressed the potential hard and soft costs associated with the project and how they could escalate depending on what [the Stanley Center] wanted to achieve. That was really the beginning of the conversation."

TIM SCHROEDER
President, Neumann Monson Architects

a sense of collective ownership" in the project, whichever form it might take.

Invoking a clever double entendre, Obernolte affectionately refers to the Stanley Center as a "green client," owing to their eco-conscious values as well as their relative newness to the green building process. "They were green to the design process and to owning a building and operating it" year-round.

The pre-design phase, beginning in early 2020, laid out in detail what those processes would look like and how a construction schedule would play out. All of this was prior to the Stanley Center even finalizing a site for their new home, let alone deciding if their eventual course would involve new construction or renovation. And of course, little did they or anyone know that it was just a matter of days before the world would come to a screeching halt.

On March 4, 2020, Stanley Center staff participated in an all-day visioning workshop facilitated by Neumann Monson. This kick-off meeting was composed of four sessions, each covering a separate set of tasks. The first, titled "Ways We Work," was key to understanding how (and where) Stanley Center staff spend their time while in an office setting, and to determine what types of workplace conditions are conducive to their culture. This first session also happened to take place at the headquarters of HNI Corporation (also founded by Max Stanley, which was soon after brought on as the project's furniture contractor), where company leaders delivered a presentation on what impact furniture can have on the work environment across all phases of whole-person wellness and employee productivity.

The all-day workshop's subsequent sessions were "Defining Goals" (what does a successful workplace accomplish, and how specifically could it fail?); "Sustainability" (what does sustainability mean for this project, and what kind of green building certification is appropriate for what the Stanley Center wants to accomplish?); and finally, "Site Analysis" (what is the site criteria for the Stanley Center's new home and how should individual factors be prioritized?).

Designating the analysis of site criteria was purposefully left to the end so that Center employees could apply their decided preferences, goals, and new-found knowledge of sustainability principles, and apply these through the filter of their needs versus their wants. For instance, how important is interior daylighting and air quality versus having proximity to the river? How important is having visibility within the community versus having adequate parking? And how important is minimizing lifecycle costs versus minimizing upfront costs?

Other workshops followed. The combined Neumann Monson and Stanley Center team were starting to develop a clearer understanding of what it was they were seeking. They were beginning to strategize on what the larger project team might look like. They even managed to squeeze in a few site and building tours. "Then, by March 13 or 14," Porter recalls, "we told people to stay home for maybe a couple of weeks."

In an instant, everything went virtual, and for a period that, of course, endured for more than a "couple of weeks." Supply chains were disrupted, businesses shuttered, entire markets and economies suddenly found themselves on tenuous ground. The Covid-19 pandemic was beginning. In theory, maybe it wasn't the best time to embark on finding a new headquarters.

SUSTAINABLE POTENTIAL

The visioning workshops gave participants the rare opportunity to explore how the Stanley Center's core values align with the principles of sustainable building design.

Mitigating anthropogenic global warming may have been key to the organization's mission, but following the workshops, Stanley Center staff began to see sustainable design practices as parts of a larger whole, as imperatives to addressing human health and wellness, conservation of water and natural resources, biodiversity, green economies, and equitable communities.

Eight guiding principles for designing the building emerged from the workshops:

1. Align design approaches to organizational mission, vision, and values, and the new brand.
2. Prioritize sustainable principles in all decision making and operations.
3. Accommodate occupant comfort, well-being, and safety to facilitate a healthy and productive workplace.
4. Inspire and empower our staff and future talent to do their best work in a positive, inclusive environment that feels like home.
5. Foster a unifying team environment and opportunities for creativity by enabling chance encounters and integrating collaborative work options.
6. Provide flexible and adaptive space choices that respect, optimize, and balance all the different ways we work now and into the future.
7. Incorporate interior and exterior amenities that support staff physically, mentally, and socially.
8. Project planning decisions will be pragmatic, financially responsible, and deliver the best value for the longevity of the organization.

"There was consensus to move forward with the Living Building Challenge," Obernolte says. "It was the more holistic direction to live out their vision and core values. It went above and beyond. And to Stanley Center's credit, they kept pushing for full Living Building certification. We just didn't have a site yet."

Using the eight guiding principles as a road map, sixteen sites throughout Muscatine were identified as possible locations for the new headquarters, eleven of which were either concentrated in the historic commercial district and/or near the river. Several were quickly discounted owing to site constraints and flexibility, proximity to downtown, and lack of ownership of the space. But the largest driving factor was each site's sustainable potential. These and other criteria were then plugged into a site matrix and scored on a scale of 1 to 10. Four sites made the shortlist — two renovations and two potential new builds.

Interestingly, several of the sites, including a few that failed to make the shortlist, ended up scoring higher on the matrix compared to the former Musser Library at Third and Iowa, the future home of the Stanley Center. But other intangibles were in play, and as the site analysis process began venturing into more granular discussions about each site's sustainable potential, the appeal of constructing a new building started to fade.

With Neumann Monson now firmly behind the wheel, their design team proceeded — by necessity — to lead the next and decisive phases of site analysis and pre-design in an entirely virtual setting, using the digital collaboration platform Mural to examine site ecology and context, solar analysis, embodied carbon, building materials, and much more; all a first for a Living Building! (Mural continued to be an invaluable tool as the integrated design team grew, trade partners were brought on board, and the project entered its schematic design and design development phases.)

Through it all, something about the former Musser Library building stuck with Porter and the Stanley Center team. Granted, the two undeveloped sites on the short list — one situated a mile from downtown and the other standing at a gateway intersection for the downtown — offered things like a generous footprint for one level of workspace (former),

SITE SELECTION

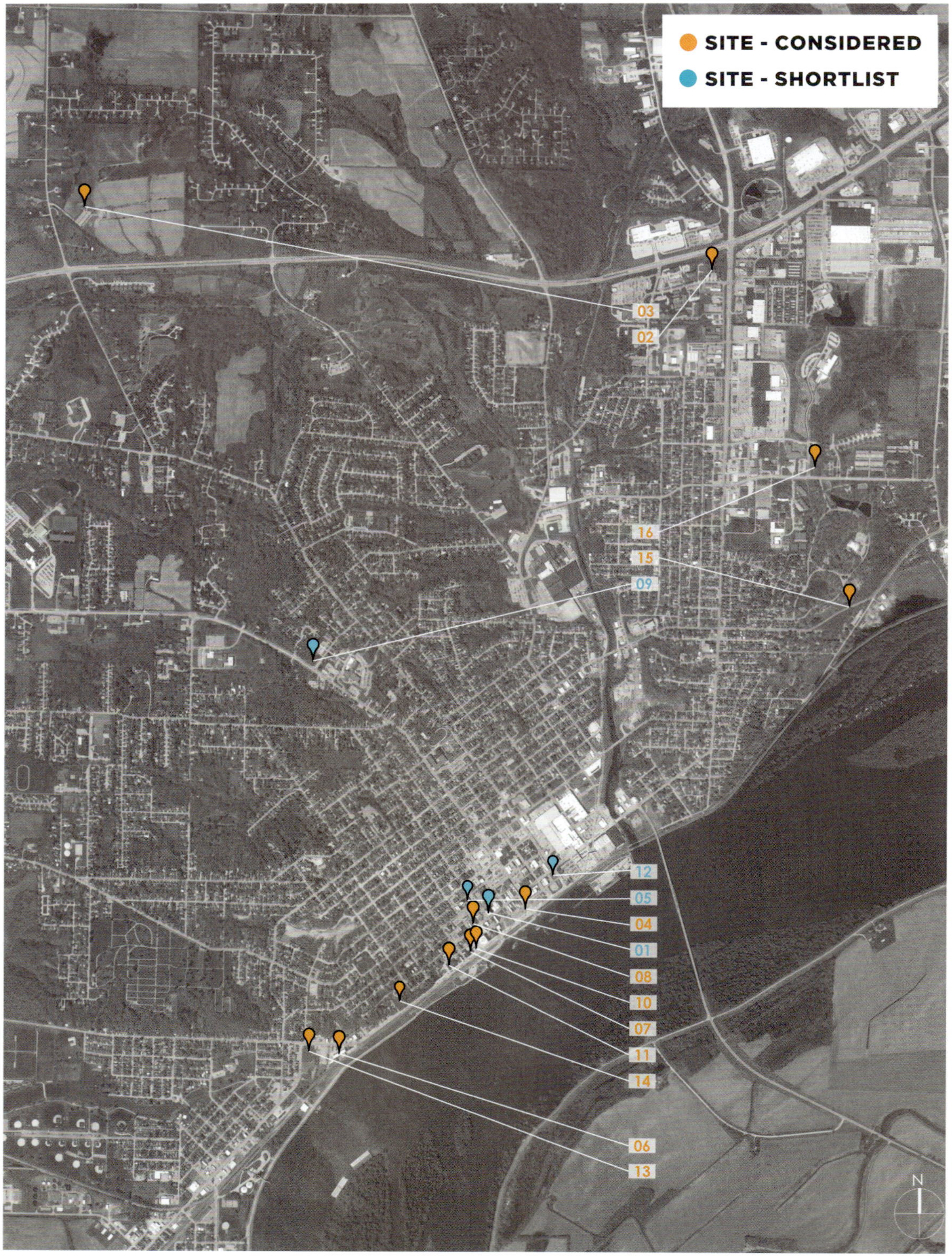

Map of considered sites throughout Muscatine; the Musser Library is site #5

SITE SHORT LIST

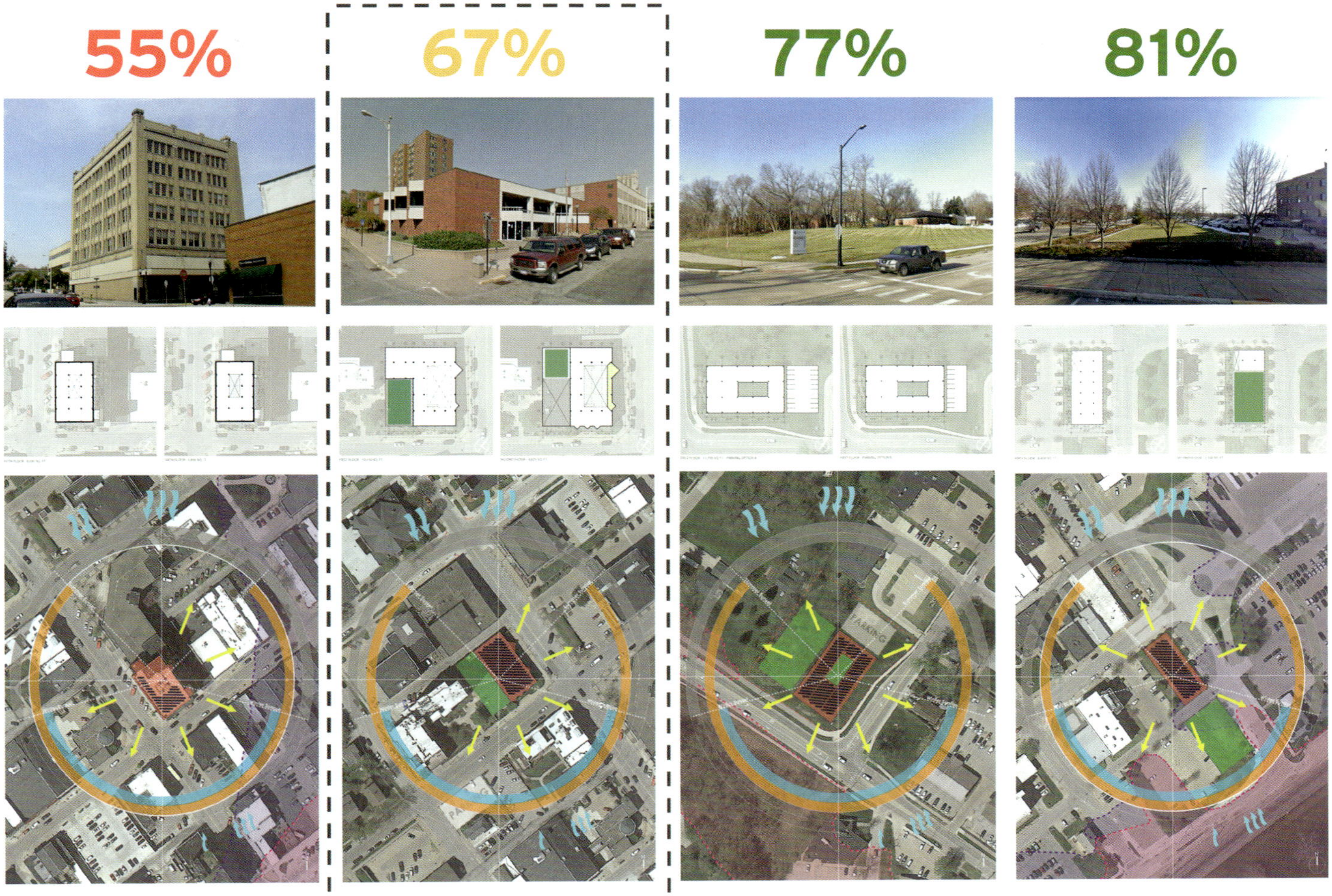

RENOVATION
More Constraints

VS.

NEW BUILD
More Control

IMPACTS
- Sustainable Potential
- Roof Decks/Gardens
- Flexibility
- Daylight & Views
- Connection to Nature
- Parking

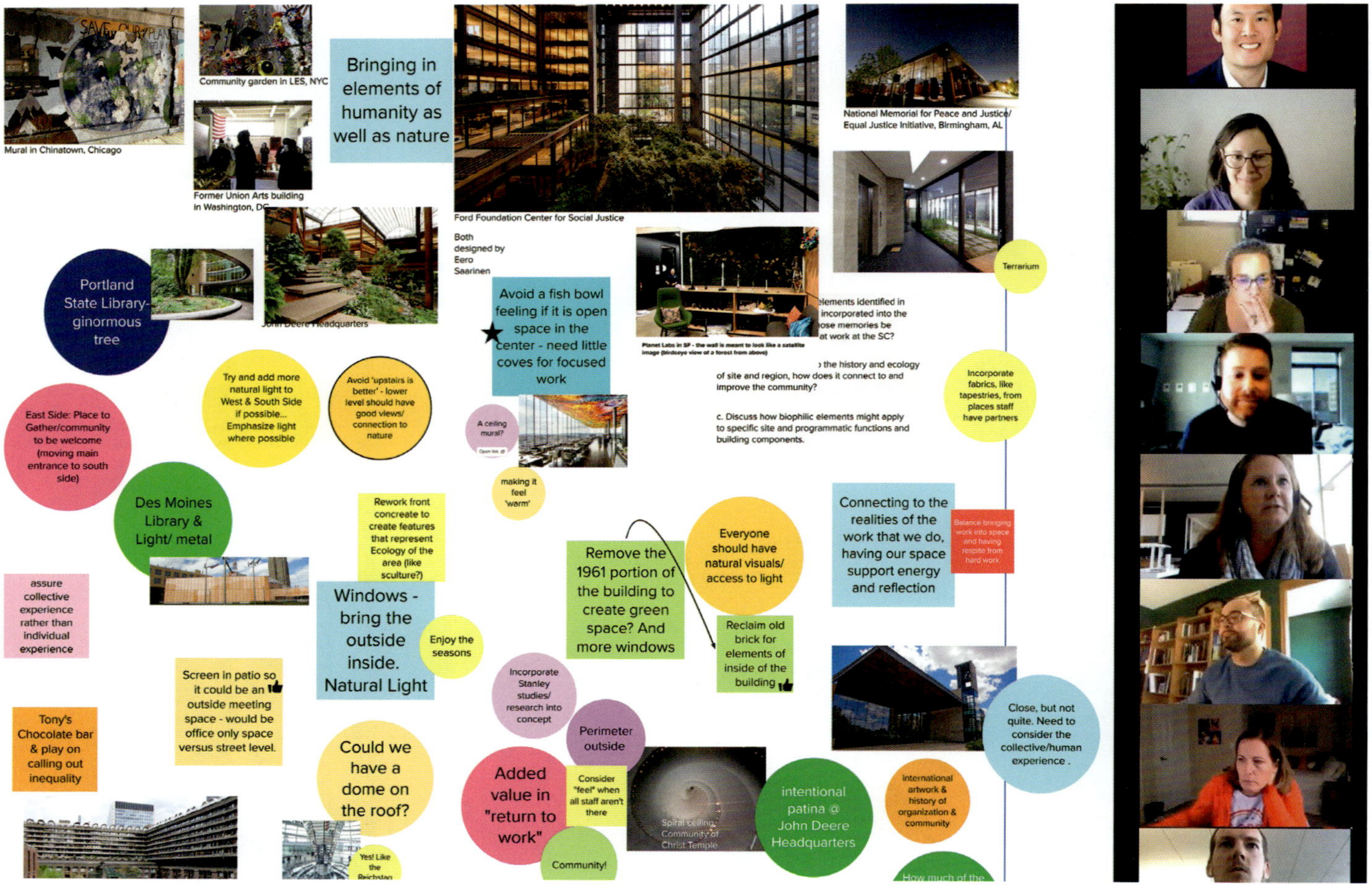

The design team conducting a Biophilia Workshop via Mural, which became an essential collaboration tool during the Covid-19 pandemic

proximity to the river (latter), and good visibility (both). But the embodied impacts of building on undeveloped land began to make the Stanley Center team uneasy.

Apart from the former Musser Library, the other possible renovation happened to be the Stanley Center's current home on the top two floors of the Laurel Building, which required a lease agreement and had major constraints related to the Living Building Challenge. However, the decision to purchase the former Musser Library and renovate it into a full Living Building wasn't just a process of elimination. The Stanley Center's heightened commitment to sustainability, and to doing good versus doing less harm, led them to the conclusion that repurposing an existing building was the right call.

"Pursuing the Living Building Challenge has had a ripple effect on our partners around the world. This is very much a way to not just live our values, but to contribute directly to the kind of change we want to see in the world in terms of combating climate change."

JENNIFER SMYSER
Vice President and Director of Policy Programming Strategy, Stanley Center for Peace and Security

"There was consensus to move forward with the Living Building Challenge. It was the more holistic direction to live out their vision and core values. It went above and beyond. And to Stanley Center's credit, they kept pushing for full Living Building certification. We just didn't have a site yet."

SALLY OBERNOLTE
Principal, Neumann Monson Architects

"As we designed and detailed the building, the Stanley Center team wrestled with cost escalations and what it would take to fully achieve the Living Building Challenge. They balanced those considerations with their values and didn't compromise on a single one."

LYNDLEY KENT
Principal, Neumann Monson Architects

The former library's open floor plan was appealing, but natural light was lacking

ANALYZING THE MUSSER LIBRARY

"You could sense its potential. It had been sitting empty for many years," says Jennifer Smyser, Stanley Center's director of policy programming strategy. "We could have built something new, or we could have opted to revitalize a part of Muscatine. The second choice aligned more with our values."

Smyser cites the Living Building Challenge's inclusion of values in its performance standards as a critical factor in informing the organization's decision to select the former Musser Library site for their future home. The larger team had also come to appreciate how those standards, or Petals, transcended building performance and had been designed to impact entire communities.

For starters, the building's orientation was primed for a large solar array. Modifying the structure to receive generous daylight and support biophilic design would be a challenge but not prohibitive. And introducing an urban agriculture program, while likewise challenging (to the point of removing a portion of the building itself), yielded an urban oasis for employees and abundant produce for the neighboring Muscatine Center for Social Action (MCSA), a local food pantry and shelter for those experiencing homelessness. These and other considerations are part and parcel of meeting the Living Building Challenge's high standards.

The choice to renovate versus build anew was one based on reducing embodied carbon and lifecycle costs. It was also an investment in Muscatine itself. At the end of the day, being truly sustainable, responsible, efficient, and equitable on a scale worthy of a Living Building all comes down to being empathetic and a good neighbor. It's about creating a building with real social impact.

It is safe to assume that, had the Stanley Center not come along when it did, the former Musser Library building on the corner of Third and Iowa would have done one of two things: sat vacant for many more years to come, or been retrofitted into something that provided little benefit beyond its walls. Solid as the structure was, without a significant adaptive reuse investment, the existing building met few practical needs of the average commercial tenant. It was too poorly lit for a traditional office space, too big and fortified for a retail destination, and too small to become a school. "It could have been something else, but probably not a big value add for the community," Porter says.

"This building was celebrated as a center of learning for over 100 years. And the building's place in people's hearts has become increasingly clear, especially as construction continued and people showed interest in the process. Continuing this location as a center of learning, through our local programming, was important to us."

MARK SEAMAN
Vice President and Director of Communications,
Stanley Center for Peace and Security

"You could sense its potential. It had been sitting empty for many years," says Stanley Center for Peace and Security's vice president and director of policy programming strategy Jennifer Smyser. "We could have built something new, or we could opt to revitalize a part of Muscatine. The second choice aligned more with our values."

JENNIFER SMYSER
Vice President and Director of Policy Programming Strategy,
Stanley Center for Peace and Security

Stanley Center's vice president and director of communications Mark Seaman considers the value add in terms of the institution's and site's history. "This building was celebrated as a center of learning for over 100 years. And the building's place in people's hearts has become increasingly clear, especially as construction continued and people showed interest in the process. Continuing this location as a center of learning, through our local programming, was important to us."

At some point, values-driven mission statements must make room for the ledger. Pursuing the Living Building Challenge requires the participation of multiple stakeholders who are wise to the realities of achieving this stringent and finicky performance standard. And those realities often involve extremely thin profit margins. Innovative projects can involve longer construction schedules and delays due to supply chain issues, or simply the need to vet a particular building material, fixture, or finish. Once the practical components and financial strains of realizing a fully certified Living Building come to light, that's when values are truly put to the test.

Musser Library exterior

Musser Library interior

This former storage area at the back of the building is now part of the courtyard; the raised area remains, accounting for the creek bed

"We referred to it as the 'Living Building Challenge puzzle,' as in, how are we going to accomplish all of the LBC's requirements on this tight urban site?"

SALLY OBERNOLTE
Principal, Neumann Monson Architects

Graham Construction and Stanley Center team members tour the building as the retrofit commences

CREATING OWNERSHIP

In the summer of 2020, Neumann Monson Architects brought on Des Moines-based Graham Construction as the project's general contractor.

Drawing from the firms' shared experience on previous projects, having worked together most notably on the GPC (Grain Processing Center) Conference Center in Muscatine, the design architect was fully aware of Graham's capacity to tackle unforeseen challenges. To Neumann Monson's thinking, Graham would be well equipped to take on the burden of working with trade partners who may be unfamiliar with the standards and rigorous requirements of handling a Living Building project.

With a Living Building, all building and construction materials must be vetted and meticulously recorded. Dumping waste materials is not an option, nor is making a last-minute run to the hardware store for a quick replacement part. The construction schedule for a Living Building is a calibrated procedure; there is room for flexibility, per se, but not improvisation. A Living Building is, above all, an educational process, wherein all stakeholders lie somewhere on the learning curve.

"There aren't a lot of materials out there that meet the standards of the Living Building Challenge," says Brian Mrozek, project manager with Graham. "To go find those materials and then procure them, there are a lot of added difficulties with that."

The Stanley Center's front entrance along Iowa Ave.

If there is one change order that affects one supplier or tradesperson, it potentially affects one or more Petals, and thus creates a ripple effect in the entire design process. With any given Living Building, such occurrences are likely. Coupled with making a Living Building during a global pandemic, disruptions become inevitable (and likely frequent). To ensure streamlined troubleshooting protocols, it is imperative that all design partners and subcontractors on the project accept accountability and are readily available to do their part.

With this reality in mind, the growing integrated design team embarked on a journey of discovery and self-reflection. This journey involved a thorough examination of multiple interrelated standards and best practices, including all seven Living Building Challenge Performance Petals, the AIA (American Institute of Architects) Framework for Design Excellence, and the practice of Biophilic Design and its ecological and ethical imperatives, as defined by Stephen R. Kellert. The team also engaged in various workshops related

"To really create a sense of ownership on a project like this and ensure operational success into the future, you need the involvement of everyone."

SALLY OBERNOLTE
Principal, Neumann Monson Architects

"There aren't a lot of materials out there that meet the standards of the Living Building Challenge. To go find those materials and then procure them, there are a lot of added difficulties with that."

BRIAN MROZEK
Project Manager, Graham Construction

to water use and conservation, space planning, and the history and present-day conditions of Muscatine's bioregion.

"We referred to it as the 'Living Building Challenge puzzle,' as in, how are we going to accomplish all of the LBC's requirements on this tight urban site?" asks Neumann Monson's Sally Obernolte. The puzzle pieces in question included how to achieve the urban agriculture component, how to infuse the building with natural light, how to harvest and store stormwater, and how to meet the building's energy demands with rooftop solar, among other concerns. But, as the scope of the project evolved and more trade partners were brought on to solve problems and tackle new ones, the puzzle metaphor ceased to be apt. Puzzle pieces don't change size, and even in disparate form, a puzzle's composition is constant. Not so much when transforming a decades-old, quasi-Brutalist municipal building into a Living Building.

Case in point, to achieve the urban agriculture component, a significant portion of the building's footprint was removed to make space for the courtyard and garden, which allowed more daylight to enter the building's interior but also lessened the available square footage for installing rooftop solar. Solving this issue, among countless others, required the attention and creativity of all stakeholders, not just a few select parties. Whatever the challenge, however big or small, the expectation from start to finish was that everyone be made aware and have a say in solving the problem at hand.

The impacts of every decision made for a Living Building project are interrelated. Based on that reality, it's helpful for every design team member, every trade partner, every sub-contractor, and every supplier to know what it is they don't know. If this is everyone's starting point, it's a good place to be — humble, curious, and thirsty. The journey of realizing the building is a constant learning opportunity; it's an opportunity to "disseminate knowledge," according to Neumann Monson Principal Khalid Khan.

By the winter of 2021, design development had begun in earnest and the project's engineering team had been selected, further rounding out the core group of Neumann Monson, Graham, and HNI. Cedar Rapids-based Design Engineers was hired as the MEPT (mechanical, electrical, plumbing, and technology) engineer; Des Moines-based Raker Rhodes as structural engineer; Massachusetts-based Integrated Eco Strategy for materials consulting; Florida-based Environmental Consulting & Technology (ECT) for civil and landscape architecture; and Baltimore, Maryland-based Biohabitats for water design strategies.

Like any worthwhile endeavor, adding more expertise and creative minds to the mix will make for a more dynamic project. It also increases the number of "chain reactions," as Khan calls them. For every design partner and contractor on the project, they are having to deal with various suppliers and manufacturers, all of whom are being asked to change (or at minimum, drastically adjust) how they do business in some respect to accommodate the standards of the Living Building Challenge, which in effect transforms the entire design process for the better, the entire construction administration process for the better, and raises industry standards across the bar. The building itself becomes an organic and constant learning opportunity for all who touch it.

"It's vital that in all of our work, in all of our travels, that we're conscious not only of the land and our impact on the land, but the history of that land and the people of the land."

DEANA DARTT
Independent Museum Consultant and Curator at Live Oak Consulting

"The Stanley Center is now visible in a way that it's never been before. This building presents so many opportunities for the Center to be more effective as a community member."

LYNNE E. STANLEY
Chair, Board of Directors, Stanley Center for Peace and Security

GROUND HEALING

To refer to something as "groundbreaking" is a celebration of its innovation and ingenuity, or at the very least its uniqueness. And as any architect or engineer knows, it is also a ubiquitous term for every time a mayor, a real estate developer, and a small group of other invited guests adorn hard hats and place ceremonial shovels into the earth to mark the start of construction. Another way to interpret it is to understand the act of breaking as inflicting harm, at which point its natural counterpart is to perform acts of healing. This is more than just wordplay; there is profound significance in adjusting the language.

On September 17, 2021, the Stanley Center for Peace and Security, along with its design and trade partners, performed a "ground healing" ceremony at 304 Iowa Avenue. In actuality, the occasion was no different than most other events of its kind. Old foundation was literally broken, partitions were removed, old finishings were stripped. But the connotations of the revised name are important.

Ground healing acknowledges and, in many respects, pays deference to the importance of this place and its history — the history of the Meskwaki people on this land, the history of U.S. settlement, of commerce and development, trade and expansion, industry and business. "We should celebrate the industrial and entrepreneurial history of Muscatine, but those things have also caused harm in our community," says Krista Regennitter, Stanley Center's program officer for global education. "Healing cannot undo past harms, but I think our building can be part of the healing process and create a space in Muscatine that will be an example for people."

Of course, there is poetic justice in an organization fostering a mechanism for healing when the wound in question is, in many respects, what has allowed that organization to thrive and carry out its work for several generations. That notion is not lost on the staff of the Stanley Center, nor is it lost on their design partners and various other collaborators who are working diligently to realize one of the world's most sustainable, high-performance buildings in this modest, off-the-beaten-path Midwest town.

But as we established throughout Part I of this book, Muscatine is no ordinary town. It's a city of profound historical importance, having produced international leaders in trade, industry, civil rights, and global philanthropy. Quiet and quaint as it may seem, from early settlement through present day, Muscatine's example carries itself with the rhythm not of a gentle thrum, but a strong and steady heartbeat, now made only stronger with the presence of the Stanley Center for Peace and Security's new headquarters, at the corner of Third and Iowa.

"We should celebrate the industrial and entrepreneurial history of Muscatine, but those things have also caused harm in our community. Healing cannot undo past harms, but I think our building can be part of the healing process and create a space in Muscatine that will be an example for people."

KRISTA REGENNITTER
Program Officer, Global Education, Stanley Center for Peace and Security

Stanley Center staff and governance members with then-mayor Diana Broderson planting a tree commemorate the "ground healing" for their new headquarters

"Neumann Monson took some really creative approaches to bringing us together as a group. We learned from working with them and having them try some of these [collaborative] approaches that they'd never tried before Covid. That became part of our learning process as a team. For example, how do you take on a project like this when, for a long time, none of us can even stand together in the same space? What does that mean for the project?"

JENNIFER SMYSER
Vice President and Director of Policy Programming Strategy, Stanley Center for Peace and Security

Top:
Team members from Biohabitats, the Stanley Center, and Neumann Monson taken at the International Living Future Institute's Living Future '23 in Washington, DC, May 2023

Bottom:
Team members from Design Engineers, Neumann Monson, and Graham Construction taken at the Stanley Center's Ground Healing Ceremony in Muscatine, September 2021

PROJECT TEAM

NEUMANN MONSON ARCHITECTS
Design Architect / Architect of Record

GRAHAM CONSTRUCTION
General Contractor

DESIGN ENGINEERS
MEPT Engineer

RAKER RHODES
Structural Engineer

BIOHABITATS
Water Systems Engineer

ENVIRONMENTAL CONSULTING & TECHNOLOGY (ECT)
Civil and Landscape Engineer

INTEGRATED ECO STRATEGY (IES)
Materials Consultant

HNI CORPORATION
Workplace Furnishings

STANLEY CENTER FOR PEACE AND SECURITY
Owner

NEUMANN MONSON
GRAHAM
Building what's important

PART III

Forging A New Home

PETALS, IMPERATIVES, AND PERFORMANCE

The Living Building Challenge Petals are designed to address more than building performance; they exist to challenge project owners and design teams to achieve symbiotic relationships between the built environment and natural systems. They are designed to realign values.

More tangibly, the Petals' intent is to foster relationships with conscientious industries and manufacturing trades that, through education and hard-won culture change, contribute to ecologically restorative, non-toxic, equitable, resilient, transparent, and humane economies.

The Living Building Challenge Petals are about creating what Neumann Monson's Khalid Khan calls "chain reactions."

The Petals themselves – Place, Water, Energy, Health + Happiness, Materials, Equity, and Beauty – are aptly named. The metaphor is unmistakable, and the Living Building Challenge's founder Jason F. McLennan knew as much. The petals of a flower are in balance with nature, their form and color specially designed to sustain life and attract pollinators to perpetuate constant cycles of new life. Some flowers, of course, have fused petals, making for a more resilient organism that can withstand harsh conditions.

The Living Building Challenge's Petals are fused. This fusion is achieved through the 20 individual yet interrelated Imperatives that work in tandem to create a wholly self-sufficient and regenerative building.

The determining factors of a Living Building's Ecology of Place (Place) – i.e., the site's ecological conditions and the context of its native bioregion – will inevitably inform how the project manages Responsible Water Use (Water) and achieves suitable Access to Nature (Health + Happiness). Likewise, a Living Building's ability to achieve Net-Positive Carbon (Energy) is complemented by the rigorous feat of achieving Responsible Sourcing and Net-Positive Waste (Materials). There are no shortcuts; a Living Building is not an either/or proposition. Where one Imperative goes, the others must follow.

This correspondence all speaks to the unparalleled rigor of the Living Building Challenge and each Petal and Imperative that compose the building blocks of this distinct achievement. The ability to grow food on-site (Place Petal, Urban Agriculture Imperative), for example, could easily be reduced to something with minimal impact and with little regard for the nature, methodology, or intent of such an operation. And the relationship between the Living Building Challenge and global building products markets (Materials Petal, Responsible Materials Imperative) could just as well be seen as incidental, and leveraged for good – sustainable extraction, responsible waste diversion, and supporting local industry – only when convenient for the owner.

But in each case, as with every other Imperative, what is baseline for the Living Building Challenge is a cut above other building performance standards and sustainable certification programs. Full transparency is baseline. The maturation and evolution of a project's native habitat is baseline. Indoor environmental conditions that actively promote occupant health and well-being are baseline. And a project's capacity for providing direct support to local businesses and economies is baseline.

Part III of this book will explore each of the seven performance Petals and, to varying degrees, the associated Imperatives as they relate to the Stanley Center for Peace and Security's future Living Building. These chapters will likewise explore the unique contributions of the various designers, engineers, contractors, manufacturers, and suppliers, and how those contributions came together to achieve each Petal and abide by the rigor of each Imperative. Finally, Part III will further examine the "chain reactions" that resulted from various stakeholders and trade professionals coming together, finding their respective places on the Living Building Challenge learning curve, and undergoing the arduous and iterative task of making a building that is responsive, sustainable, regenerative, restorative, inclusive, beautiful, inspiring, and net-positive on all scales.

Stanley Center
FOR PEACE AND SECURITY

FLOOR PLAN:
FIRST FLOOR

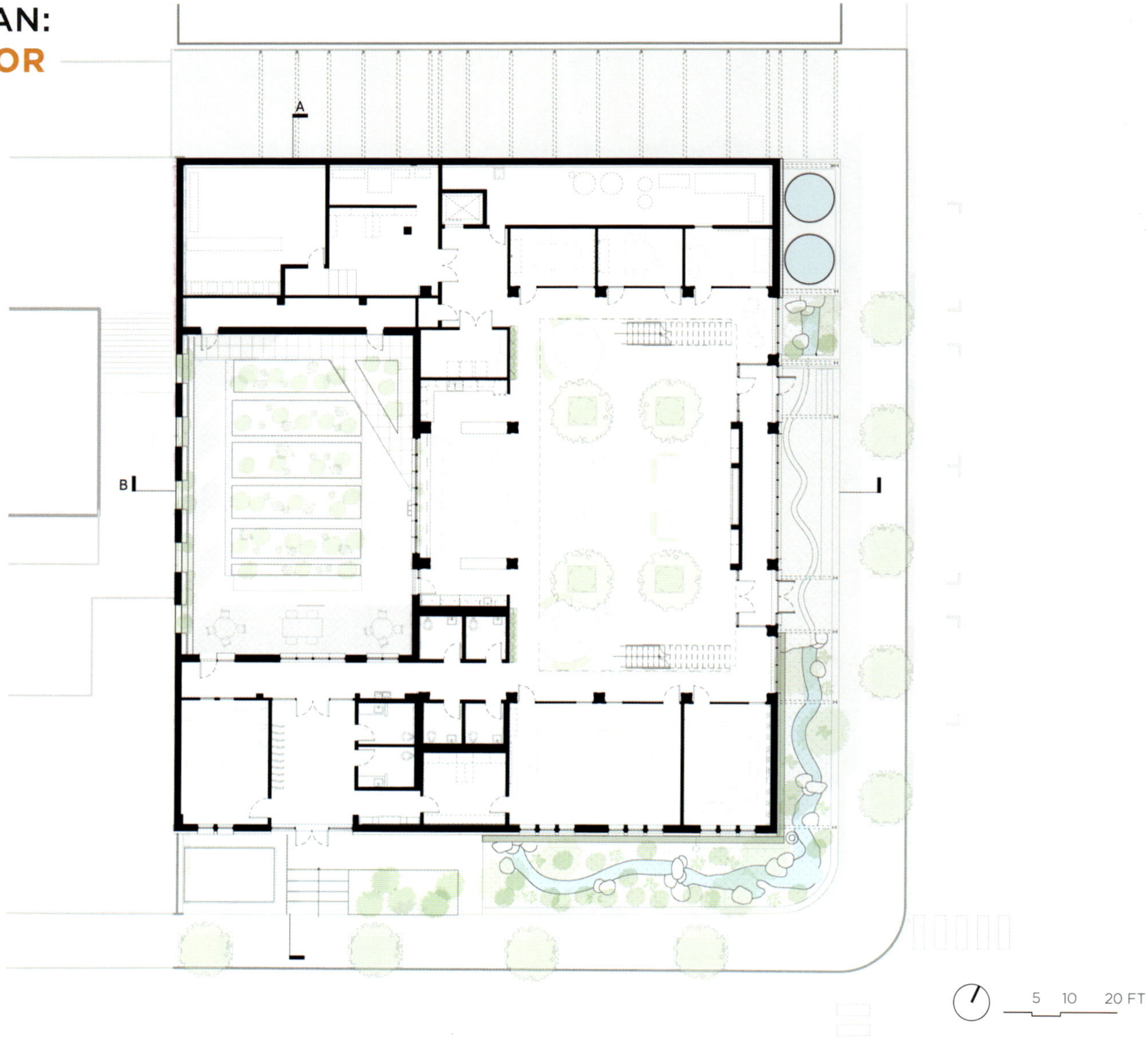

FLOOR PLAN:
SECOND FLOOR

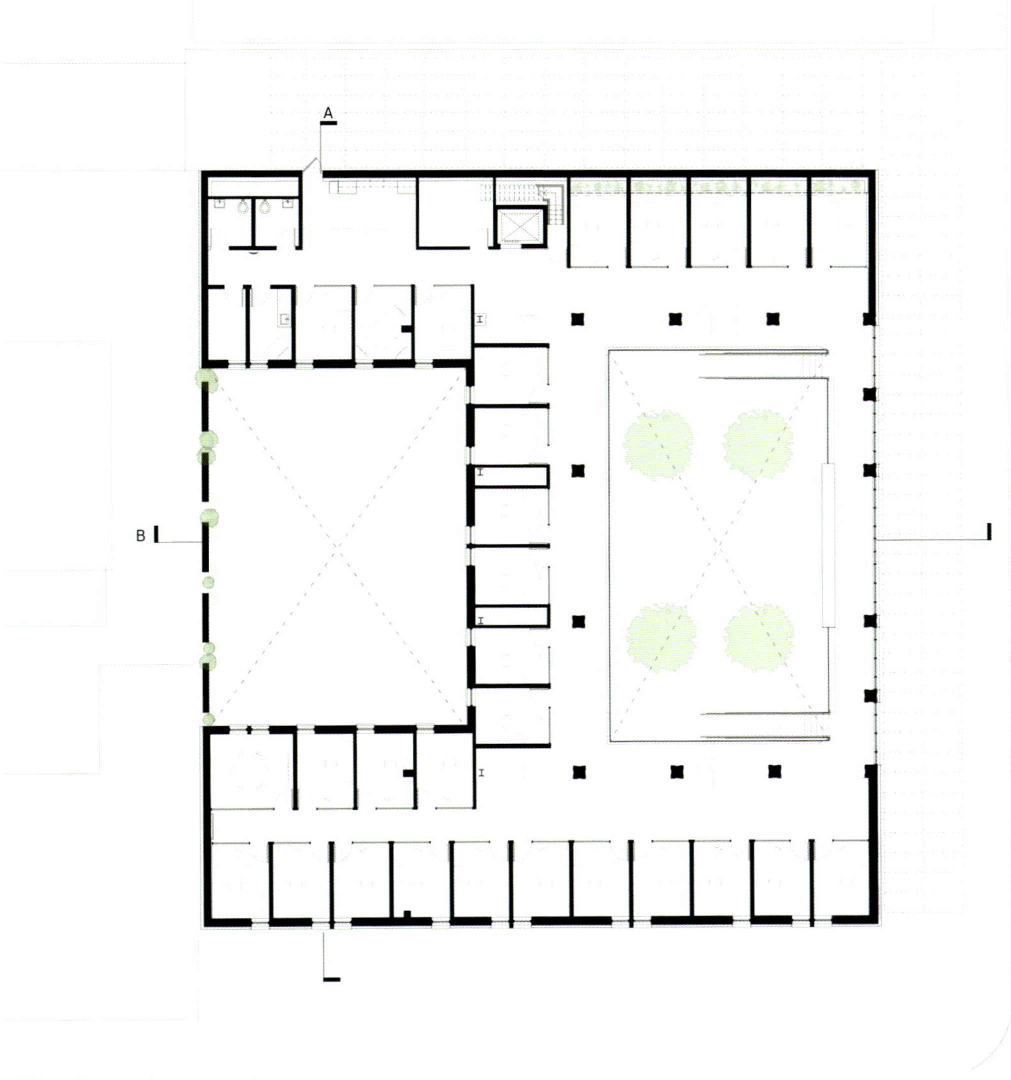

THE PLACE PETAL

Referencing Place and History

Downtown Muscatine in the early 20th century and the early 21st century

"The Stanley Center's reasons for staying in Muscatine helped us understand this town and its importance. We had the opportunity to explore Muscatine and its history, which guided decision-making when looking at site options and helped us find a location that would align with the Living Building Challenge. Their connection to Muscatine underpinned everything we wanted to amplify in the project and played into the Ecology of Place."

KHALID KHAN
Principal, Neumann Monson Architects

REFERENCING PLACE AND HISTORY

Once the former Musser Library on the corner of Third and Iowa had been selected and purchased by the Stanley Center, certain realities started to settle in.

For starters, the Center had never worked with an architect; since its founding, in 1956, the organization had operated out of a series of leased office spaces and made do with what they provided. So, acclimating to the lexicon of design and sustainability was important, as was adjusting to the rhythms and inherent challenges of creating a highly customized built environment.

More so, it was immediately evident that what the Stanley Center was about to embark on involved much more than the adaptive reuse of a building. It involved, according to Keith Porter, "doing a deep dive into this piece of ground." This analysis included looking at the land and understanding the conditions of its natural state, and what the lives and livelihoods of the people who once called this land home, pre-settlement, looked like.

Part of the intent of the Living Building Challenge's Place Petal is to question "how to protect and restore a place once it has been developed, and how to encourage the creation of communities that are once again based on the pedestrian rather than the automobile." It continues, "As previously disturbed areas are restored, the trend is reversed, and nature's functions are invited back into a healthy interface with people."

The Petal's intent "lined up really well" with the Stanley Center's organizational intent to build meaningful relationships with Indigenous communities in and around Muscatine and helped "deepen those conversations we were already having with Indigenous people in the area," Porter says. Such conversations also happened to prove critical to the project's team understanding of the site's ecological history and reference habitat, the ecosystem that naturally occurred on the site prior to "fundamental alterations" like urban development and sprawl.

The physical place that the Stanley Center now calls home has undergone various iterations. Nearly two centuries ago, the shores of Muscatine began to build out to accommodate steady industrial growth and trade, starting with lumber, then followed by the excavation of freshwater mollusks and the manufacturing of pearl buttons. Looking deeper, the Mississippi River basin, which covers an immense swath of land and riverways that spans western Montana to Pennsylvania, and from northern Minnesota on down to the Gulf of Mexico, comprises a vast range of bioregions and habitats. At the basin's approximate center, in Muscatine, the area's native landscape is upland oak savanna, or Midwestern savanna, defined by pockets of grassy woodlands, forest glades, open canopies, and a mix of oak species. These environments also contain a diverse mixture of plant life that attract pollinators, including tallgrass prairies, flowering plants, and woodland species such as oak, maple, basswood, and beech.

The mandate of creating a building that successfully references this habitat as well as — in some relatively small but significant measure — restores it, is enough to put the 'Challenge' in Living Building Challenge. (But of course, it doesn't stop there.) Integral to the larger challenge, and the Place Petal in particular, is to accomplish this end goal in such a manner that its impacts are felt well beyond the confines of the building's site. After all, a

SITE CONTEXT PLAN

1. Relocated Historic Alexander Clark House
2. Original Location of Alexander Clark House
3. Muscatine Center for Social Action (MCSA)
4. Stanley Center for Peace and Security
5. Historic Muscatine Secondary School
6. Clark House Apartments (Public Housing)

The home of Alexander Clark was nearly demolished in 1974 to make way for the Clark House affordable housing building. Before this could take place, a small group of citizens, led by Elizabeth "Bette" Veerhusen, convinced the U.S. Park Service the home was eligible for listing on the National Register. The group successfully lobbied and raised enough funds to move the home to a new site, just up the block on W. Third St.

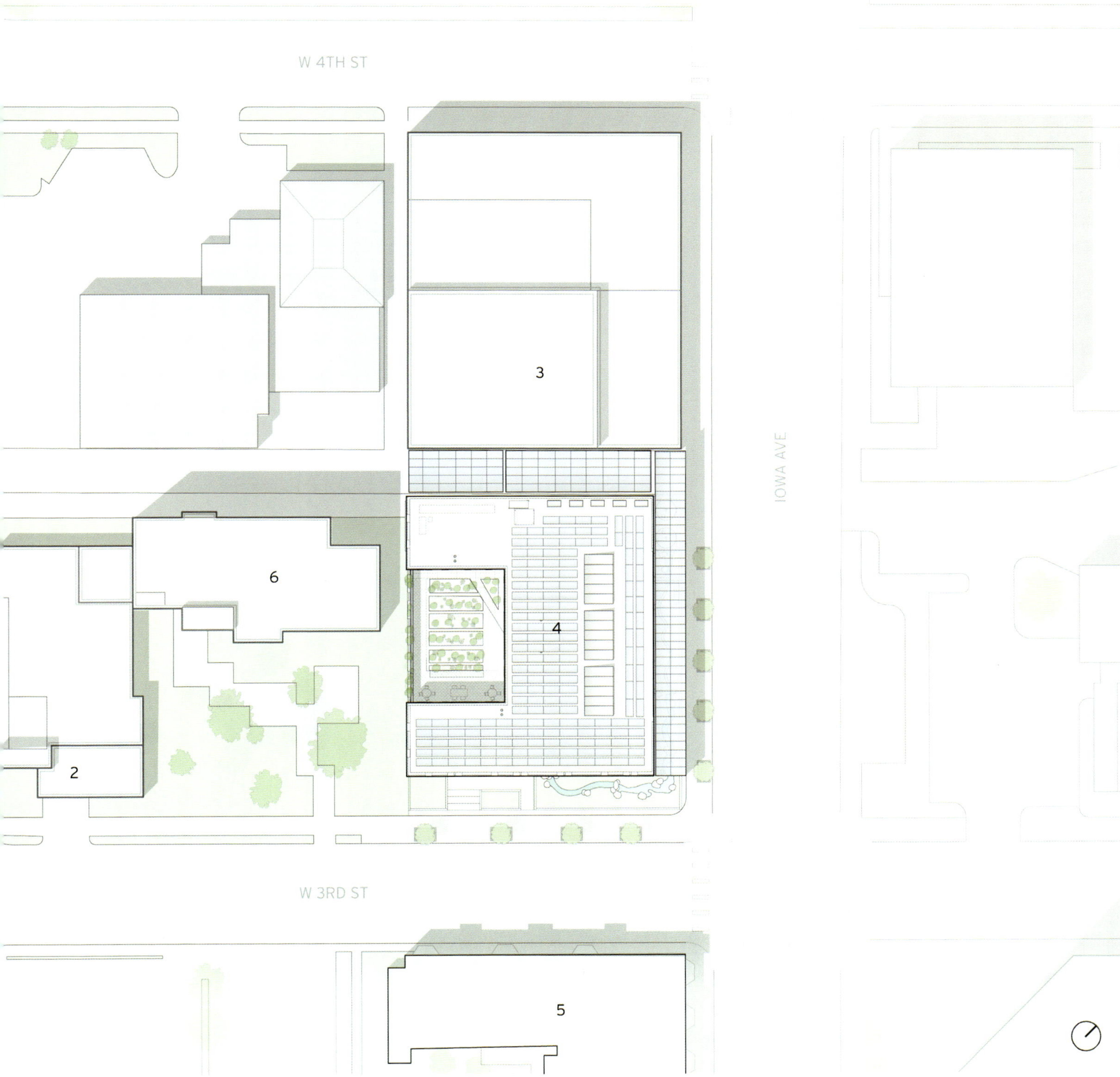
W 4TH ST
3
IOWA AVE
6
4
2
W 3RD ST
5

"Our Biophilia Workshop helped us focus on the Place Petal and root this project in its location. Learning more about the general ecology of the area, the Mississippi River Basin, and the history of Muscatine was invaluable during design."

LYNDLEY KENT
Principal, Neumann Monson Architects

sustainable building, by definition, merely sustains itself and the basic needs of its occupants. A restorative building, on the other hand, yields net benefits for the surrounding community.

In many respects, this is an oversimplification, and is meant in no way to cast aspersions on the merits of sustainable design. On the contrary, sustainable design principles are key to achieving broader holistic measures found in examples of regenerative building, restorative development, and circular economies. And that is just the case here.

The Stanley Center resides on a tight urban site, with modest setbacks along the corner of Third Street and Iowa Avenue. The 11-story public housing tower, known as Clark House (named after Alexander Clark), stands at Stanley Center's western edge; the non-profit Muscatine Center for Social Action (MCSA), an important civic organization that provides food, shelter, medical care, and counseling services, stands to its north, separated only by a narrow alleyway. The Alexander Clark House (not related to Clark House) is a block and a half away; Stanley Consultants' home office is across the street; various municipal buildings, private businesses, and houses of worship are within a short walk from Stanley Center's front door.

"We want to be good neighbors," Porter says. "The Living Building Challenge process led us to develop much deeper and richer relationships."

The question remains: exactly how does one "restore" and "invite back" nature's functions under such conditions? The answer, like the Living Building Challenge itself, is manifold. By focusing on restoring a place (i.e., its ecology and habitat), one naturally becomes preoccupied with related elements, such as water, materials, energy, and human wellness. Each consideration begets another, and no Petal is ever a system of isolated functions.

View of the courtyard, a space once occupied by library storage rooms and offices; the raised step seen on the lower left conceals the creek bed and historical sewer line

WELCH
APARTMENTS
EXIT

Evolution of the creek bed, circa late 19th century through present day, centered on the site of Third and Iowa

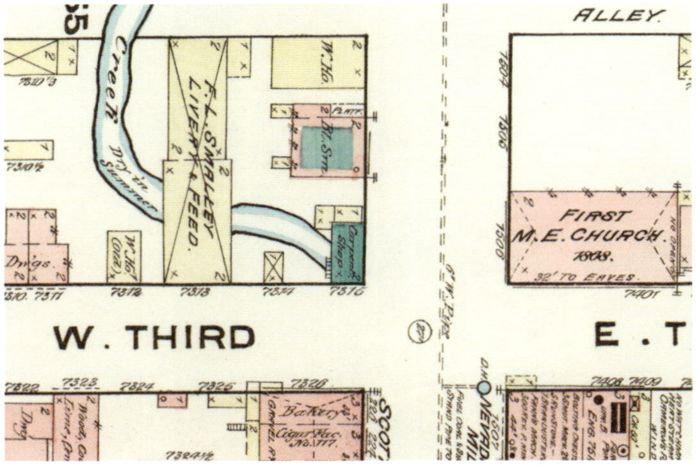

IOWA AVENUE
5 MINUTE WALK
EAST 2ND AVENUE
MISSISSIPPI RIVER
LEGEND
1 STANLEY CENTER FOR PEACE AND SECURITY
2 MUSCATNE CENTER FOR SOCIAL ACTION (MCSA)
3 CLARK HOUSE APARTMENTS (PUBLIC HOUSING)
4 HISTORIC MUSCATINE SECONDARY SCHOOL
5 HNI CORPORATION
WATER
HISTORICAL SEWER LINE
500-YEAR FLOOD LINE
MAIN ACCESS ROADS
ROADS
BUS STOPS
RESTAURANTS
CONVENIENCE STORES
GROCERY STORES
HEALTH SERVICES
RECREATION

"The building's setback allowed us some opportunity to get in some different site components and landscape. It's all highly localized. Not that these plants don't grow elsewhere, but we wanted to make sure we selected plants that would be appropriate to the local and regional landscape."

JONATHAN LAWRENCE
Senior Landscape Architect,
Environmental Consulting & Technology (ECT)

"The Place Petal impacts the entire project and touches on several other Petals, including how we treat water on the site, its connection to the creek bed, the biophilic patterning in some of our finishes, and much more."

LYNDLEY KENT
Principal, Neumann Monson Architects

"It was a pretty stout building and didn't provide views of the river, which was something the Stanley Center had expressed a preference for. But when we walked into the space, we saw the potential to do something unique in the atrium, even if we didn't know what that would be."

SALLY OBERNOLTE
Principal, Neumann Monson Architects

MEETING OF THE (BIOPHILIC) MINDS

The path to achieving the Place Petal truly began with the Biophilia Workshop. In September 2020, the entire project team to date, including members of the Stanley Center, Neumann Monson Architects, and Graham Construction, a former Musser Library librarian, an ecologist, and a local historian, convened for a two-day working session to study, explore, debate, and ultimately, find opportunities to "root the project" in this very specific location and bioregion.

"The Stanley Center deciding to stay in Muscatine was a huge factor," says Neumann Monson's Khalid Khan. "The Living Building Challenge allowed us the possibility to do this deeper dive all around town and look for opportunities." Those opportunities lay in the details.

The two-day workshop explored, among other things, social ecologist Stephen R. Kellert's six elements of biophilic design: Environmental features, Natural shapes and forms, Natural patterns and processes, Light and space, Place-based relationships, and Evolved human-nature relationships. As the late Kellert (1943-2016) surmised, understanding the overarching concepts of biophilia and biophilic design are not terribly difficult. The challenge is finding tangible applications for each inherent design element (environmental features, et al.), and successfully coalescing them within a physical building that is sustainable and continually in use. There are many Imperatives across several performance Petals that contribute to (and benefit from) a robust biophilic design program, but chief among them is the Place Petal's 'Ecology of Place.'

The ecological history and reference habitat of the site were critical factors to achieving this Imperative. "Biodiversity and seasonality were key," says Jonathan Lawrence, senior landscape architect and horticulturist with ECT (Environmental Consulting & Technology), the project's landscape architect, whose firm also provided civil/water resource engineering. Citing the reference habitat of upland oak savanna, the Stanley Center site was re-vegetated using all native plant species historically found within Muscatine County — more than 2,000 plants from 20 different families of perennials, shrubs, and trees.

This biodiverse, native landscape program was also critical for reintroducing various species of insects and pollinators that are key to perpetuating healthy fertilization cycles for the bioregion. Lawrence, ever mindful of the Living Building Challenge's Petals, sees his company's work extending beyond 'Ecology of Place.' "This supports Beauty and Health and Happiness. It promotes access to nature. Not just for the humans using the building but the wildlife as well."

The Stanley Center site also benefited from its relatively generous setbacks. Unlike most of the building's neighbors, like the MCSA just up the block or the Welsh Apartments to its south, the two-story former library was designed with some modicum of curb appeal, befitting most municipal buildings. These setbacks along its easterly and southerly public-facing sides, along Iowa Avenue and Third Street, respectively, enabled the broader landscape design team to introduce plantings that would become a pleasant and green civic asset for pedestrians in this urban context, as well as support the site's new bioretention area residing at the southeastern corner, where Iowa meets Third, the lowest elevation point for the entire lot.

With further consultation from ECT, the team looked at the site's most important yet inconspicuous ecological feature: the underground creek bed that diagonally spans the site from its northwest to southeast corners. The small creek runs "dry in summer," according to a Sanborn map from 1883, and has been enclosed within a subterranean culvert for no less than 130 years. The underground tunnel informs the undulating curves and elevational changes within the site itself.

The culvert obliged the larger design team to consider how water would flow throughout the site, be collected, and redirected for bioretention in the event of overflow (all of which is explored further in Part III's Water Petal chapters). Complementing the water flow design, the culvert's presence also inspired a "biophilic patterning" design using permeable pavers along the front entrance, which are configured as a meandering creek to mirror the motion and trajectory of this naturally occurring body of water.

Finally, it obliged the team to carefully consider the Stanley Center's urban agriculture component, which is primarily situated in the enclosed, 2,469-square-foot courtyard towards the back of the lot, bordering Clark House. Given the unusual angular ascent of the courtyard, which accommodates the aging culvert that, if exposed, would resemble a troublesome tree root, it was simply impossible to plat a standard courtyard with garden beds. What transpired, rather, was an urban agriculture program - comprising 914 square feet of the courtyard area - that adhered to the unique configurations and elevations of the site and emboldened Stanley Center's commitment to its community.

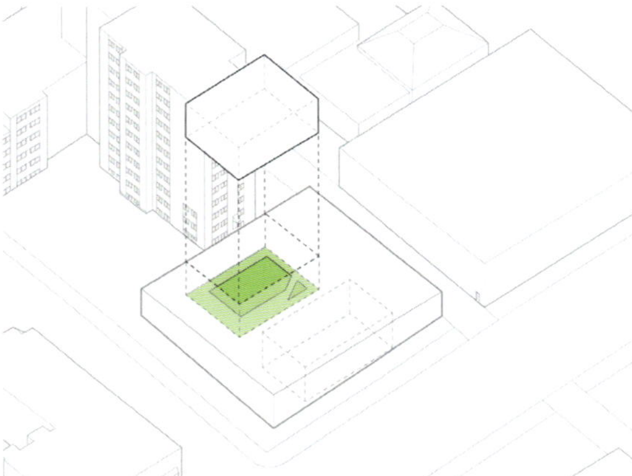

A COURTYARD FOR FOOD AND COMMUNITY

The biggest concern regarding the design and planning of the Stanley Center's urban agriculture program was location; the second concern, as is typically the case with most other critical infrastructure components in a regenerative building (e.g., HVAC systems, PV arrays, cisterns) was right-sizing.

"Being on an urban site, you have to get creative," says Neumann Monson's Lyndley Kent. Her colleague Sally Obernolte concurs. "It's not like there's green space somewhere for us to say, 'Yep, we can go grow a garden there!'"

Not unlike the team's earlier analysis of possible building sites around town, several different scenarios were considered. Place gardens on the roof? Or along the street? Neither of these options seemed viable, but each discussion proved illuminating for the larger design team regardless. When considering rooftop gardens, this called into question means of egress, weight loads and structural capacity, available square footage for solar panels, and remaining available surface area for rainwater collection. And when considering along the street, possibly with the use of evaporative greenhouses, the prospect of implementing this critical programming component out in the public realm

"Being on an urban site, you have to get creative. We took a couple of approaches and ended up with the courtyard and the lettuce walls that flank the cafe and kitchen. The courtyard helped with the Stanley Center's strategic relationship with MCSA next door, allowing them to plant things they don't always have access to."

LYNDLEY KENT
Principal, Neumann Monson Architects

Graham Construction staff assisting with unloading food pantry delivery for MCSA

"It's all interconnected. Having all that soil and weight on top of the building for what we ideally wanted to do with depths and things like that just wouldn't have worked. The other thing it would have potentially limited to some degree is collecting the rainwater. If the roof was filled up with agricultural areas and landscape, well, we're not collecting water from the landscape. We're managing it, but we're only collecting rainwater from the roof to use for the building and the site systems. That's why it was such a holistic conversation to talk about all of these different factors."

BRANDON FETTES
Intern Architect, Neumann Monson Architects

"We asked, what if we removed a portion of [the building]? And what are the advantages of doing that, if we can increase access, if we can increase daylighting into the building. Then, we now have an area that can fulfill the urban agriculture [imperative]. It can be this very private and kind of contemplative area where staff can gather. It was a pretty big idea for the whole project."

JONATHAN LAWRENCE
Senior Landscape Architect,
Environmental Consulting & Technology (ECT)

felt too unruly. Chain reactions came to bear. The Place Petal was now informing the Energy Petal, the Water Petal, the Equity Petal, and so forth. Easy answers eluded everyone.

When the solution to introduce a courtyard was decided, it paved the way for yet another slew of challenges. As ECT's Jonathan Lawrence recalls, the Stanley Center "didn't need the entire building, so we asked ourselves: 'What if we removed a portion of it?'"

Wild (and costly) as the notion may have seemed, it proved the most viable and, in the end, yielded new pathways for implementing other Petal Imperatives, like Access to Nature (Health and Happiness), Living Economy Sourcing (Materials), and Beauty and Biophilia (Beauty), among others. By altering the building's basic form, from a solid block to a 'U' shape with an enclosed courtyard, the Stanley Center could successfully line north- and south-facing exterior walls, as well as newly created interior-facing walls, with a continuous series of private offices and workspaces, and consequently bring more daylight into the building.

As for the actual courtyard, which is accessible from multiple points on the first floor, the Stanley Center's outdoor agricultural program, which utilizes more than 900 square feet of cultivated plots, is prioritizing fresh herbs, alliums (e.g., cultivated onions, garlic, scallions, leeks, chives), edible flowers, kale, mustard greens, and other lettuces. A complementary interior agricultural component includes two irrigated green walls, located near the kitchen, that grow a variety of salad greens that are available to staff. (With the interior addition, the Center's total urban agriculture program comes to 1,073 square feet of cultivated grow space.) The crop selection was the product of what Lyndley Kent calls a "strategic alignment" between the Stanley Center and the MCSA.

In early 2020, the Stanley Center formed its Office Space Steering Committee along with seven focus teams, each tasked with finding effective design and programming strategies specific to their topic. One of these topics, naturally, was Urban Agriculture. Concerns over location and orientation, right-sizing and site management, were all paramount. But in the end, these were questions of logistics, not purpose.

Left: Portions of Stanley Center's urban agriculture program, including herbs and vegetables, supplement MCSA's food pantry *Right:* Harvesting edible greens for lunch has become a daily ritual for some Stanley Center staff

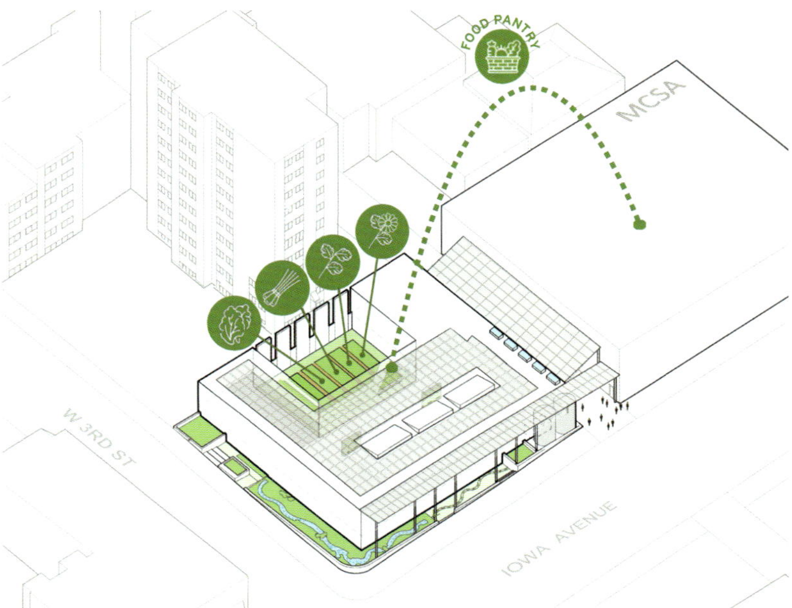

GOOD NEIGHBORS

The Muscatine Center for Social Action (MCSA) is a local non-profit established in 1991 to "meet the social needs" and "make a difference in the quality of life" in the Muscatine area, according to its website. The organization includes an on-site homeless shelter, mental health counseling services, pediatric dental care, an adult drop-in center with social workers on call, and a food pantry, all in addition to confidential off-site shelters for families and victims of domestic violence.

Once it became known that the Stanley Center would become MCSA's new neighbor, it was never a question of if these respective organizations would align their values and missions, but how. "It's a very important partnership," says Robert Albee, interim director of MCSA and a long-time volunteer with the organization. "There are so many positives to renovating the [former library] next door, and the Stanley Center is very supportive of the work that happens here."

Conversations between the Stanley Center and MCSA began shortly after the building at 304 Iowa Avenue was purchased, but urban agriculture wasn't the immediate topic. Rather, early discussions focused on, among other things, the proposed construction schedule and how the constant comings and goings of workers and equipment might interfere with MCSA's daily programming. But according to Albee, both the general contractor and Stanley Center staff were very mindful of the disturbances caused by a construction site, particularly on a tight urban footprint. Whenever deliveries were made to MCSA, Graham Construction and Stanley Center employees would drop everything (within reason) to assist with unloading the trucks. And when MCSA patrons would queue up each Tuesday and Thursday morning when the food pantry opens,

"It's a very important partnership. There are so many positives to renovating the [former library] next door, and the Stanley Center is very supportive of the work that happens here."

ROBERT ALBEE
Interim Director,
Muscatine Center for Social Action (MCSA)

"At different times of day, there might be a queue that comes out of MCSA's door and lines up in the alley. The PV canopy that stretches over the alley creates more shade in the summer and gets [their patrons] out of the elements, especially if it's raining or snowing."

LYNDLEY KENT
Principal, Neumann Monson Architects

"It was important to not only provide food to the on-site employees but also donate some of that material next door and service the community, to be a good neighbor. The urban agriculture program was able to focus on herbs and other things MCSA needed for their patrons."

BRANDON FETTES
Intern Architect,
Neumann Monson Architects

which can include up to 120 people over a two-hour period, accessibility was assured, and noise levels were kept to a minimum.

As the design of the Stanley Center took shape, the two organizations engaged in three separate but interrelated discussions that were mission critical to their continued partnership: food, water, and energy.

The first, of course, concerned the Stanley Center's Urban Agriculture program. MCSA's food pantry has been helping community members in need for decades. "There's no cost and there are no questions asked. If you need food, you can come in and get it," Albee says. But historically, fresh vegetables and other produce weren't always easy to come by. So, when it came time to design and curate the building's agricultural component, the Stanley Center simply posed the question to MCSA: 'What do you need?' The answer was herbs, alliums, and select other vegetable crops with a decent shelf life. The plan going forward, starting with the 2023 growing season, is for the Stanley Center to donate much of its outdoor harvest to MCSA's food pantry program.

The on-site harvesting and use of water also played a key role in aligning the two organizations. "The two buildings are partners in water usage. Or I should say reduced water usage," Albee says. The MCSA houses several dormitory-style rooms and other facilities for its staff, volunteers, and patrons, all of which place a premium on water demand. In replacing and upgrading the facility's water fixtures, utility costs are minimized, and the organization measurably reduces its dependence on municipal systems. (Water strategies, as well as the Stanley Center's commitment to becoming a "handprinting" organization, are explored further in Part III's Water Petal chapters.)

The Stanley Center's energy strategy, strangely enough, has also directly (and positively) impacted MCSA's food pantry operations. Every Tuesday and Thursday morning, local patrons queue up in the alley that separates the two buildings, rain or shine. (The location of the queue was temporarily moved during construction, for safety reasons.) Once the decision was made to remove a section of the Stanley Center's building to make room for the courtyard and 914-square-foot garden, it reduced the available square footage usable for rooftop solar panels. So, to meet the building's power needs, the entirety of the photovoltaic system — while accounting for skylights and one corner with extended shading — had to be expanded beyond the scope of what was available. (Energy strategies are explored further in Part III's Energy Petal chapters.)

The solution was to create canopy overhangs, one along the building's front entrance, facing Iowa Avenue, and another extending over the alley between the Stanley Center and MCSA. This innovative design and building performance solution has in turn yielded a civic asset. The canopy overhangs provide shade and shelter from the elements and introduce important placemaking principles to a public space that was until recently used strictly out of necessity. Now, it is a place for gathering and respite, and soon, a destination for public art.

"One of the important aspects of the site is there was no dedicated on-site parking. We have continued not to provide on-site parking, nor have we set aside space for future parking."

SALLY OBERNOLTE
Principal, Neumann Monson Architects

A WALKABLE NEIGHBORHOOD

To realize what the Living Building Challenge calls a "human-powered lifestyle," building owners in most contexts — urban or otherwise — must contend with the pervasive culture of the automobile. Whole cities and towns have been planned to accommodate cars and traffic; this isn't exactly news. But as more enlightened minds within urban planning circles have helped the world realize of late, auto congestion is not some inevitability that is beyond our control. If you treat the patient differently, adverse symptoms can subside.

At the Stanley Center, "everybody's encouraged to do something different to get to work or get to the building," says Neumann Monson's Sally Obernolte. Indeed, when the integrated design team was consulting with their client in the early stages of the project, in the site selection phase, the Musser Library site scored particularly high in walkability. During site analysis, on a scale of 1 to 10, the building's general proximity to adjacent amenities scored an '8', matching the site's 'Resiliency/

The dedicated staff entrance and bicycle docking room, looking out onto the courtyard

"Everybody's encouraged to do something different to get to work or get to the building."

SALLY OBERNOLTE
Principal, Neumann Monson Architects

"In our scoring matrix, the site was one of the few options within the two or three blocks in Muscatine that had a higher walkability score."

TIM SCHROEDER
President, Neumann Monson Architects

Minimized Risk' score and outscored only by 'Ownership/Control' (9) and 'Proximity to Downtown' (10). As a bonus, it's just a three-minute walk to the Mississippi River.

The opportunity to create a built environment that bolstered Muscatine's overall walkability was palpable. Of course, it would have to be. The only public parking available in the Stanley Center's immediate radius is metered on-street parking. The city's main public transit amenity, the MuscaBus, provides reasonably priced shuttle service within town. (A regular bus stop is located up the block near the entrance of MCSA.) Beyond options that require walking or shuttle bus, or carpooling for staff who reside in the farther reaches of the town or county, the Stanley Center features a dedicated and ADA-accessible room for bike storage, accessible at the building's side entrance along Third Street, which also doubles as a mud room and includes a separate adjacent room with staff lockers, restroom facilities, and showers. (This part of the building can also act as an emergency shelter, a feature that is explored further in Part III's Energy Petal chapters.)

To achieve human-scaled living, according to Neumann Monson's Lyndley Kent, "we talked extensively about how we wanted to approach that design strategy," particularly as it concerned the public realm. The canopy over the alleyway was a big piece of achieving that strategy, as were the cisterns, the exterior landscaping and sidewalk plantings, and the building's ease of accessibility along the avenue and side street. And the alley canopy also presented another opportunity to transform a transient public space into a festive one, with murals produced by local artists that will grace the side wall of the MCSA building.

The Stanley Center's design choices reflect its mission. Assuredly, the organization's staff and leaders will continue to perform "heads down" and "nose to the grindstone" work, recalling Keith Porter's words, but will do so in a facility that, by design, demonstrates purpose, equity, empathy, and a civic commitment to the community of Muscatine.

PLACE PETAL OVERVIEW

- Indigenous territory of the Meskwaki, Ogallala Sioux, and Kickapoo
- Site of the Musser Library, 1901-2018
- Across the street from the first integrated school in Iowa

Muscatine citizens gathering along Iowa Ave, in front of the newly completed Stanley Center, in 2023

THE WATER PETAL

Honoring the Watershed

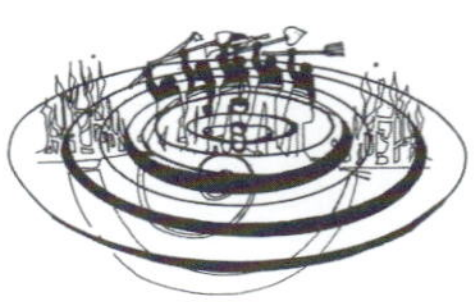

FOURTH SLOUGH-MISSISSIPPI RIVER
HUC 12 - WATERSHED
Papoose Creek
FLOOD
ZONE
EXISTING
CREEK BED
STANLEY CENTER
PROJECT SITE
Mississippi River
Mad Creek
Mad Creek

"We spent a lot of time understanding and coordinating how water moves between systems on this site. It's those little things that can make or break a project. Integration is always important for Living Buildings."

PETE MUNOZ
Senior Engineer, Biohabitats

HONORING THE WATERSHED

Challenges and constraints are just opportunities in disguise. This approach must be adopted to make any building a fully certified Living Building. But it's more than a design approach; it is a philosophy and a mantra of patience and self-education.

One of the main constraints involving the Stanley Center was the site's hydrology. The existing creek bed — encased and diverted into a storm sewer for over a century — that bifurcates the lot diagonally was "probably the most cumbersome thing we all had to work around," according to Neumann Monson's Sally Obernolte. "It's directly under the floor and the building steps up" due to the slope of the tunnel, which ascends from the southeast corner of the site, where Iowa Avenue and Third Street intersect, to the northwest corner, where the courtyard meets the alleyway.

The elevation change is hidden in plain sight; it's apparent, but subtle. The courtyard gradually steps up to account for the storm sewer. Where the change is most abrupt, an elevated slab is cater-cornered and wedged against the building, effectively creating a stage within the courtyard that overlooks the garden beds and trellised green wall at the back end of the lot. And where the tunnel's footprint meets the physical building, banquette seating along the back edge of the kitchen cleverly conceals the site's undulating footprint. Incidentally, these unique conditions also impacted how the design team configurated the building's restrooms, ensuring that any newly installed plumbing didn't interfere with the century-old storm sewer. "We have water happening on both sides of this tunnel that we cannot cross," Obernolte says.

Water and landscape are interrelated components. Adversely, when a sloped landscape and 100+ year-old urban infrastructure become interrelated as well, the constraints to achieving sustainable site performance, let alone the Living Building Challenge's Water Petal, can seem insurmountable. And in this case, they were. Which is why the team went under the tunnel.

According to Jonathan Lawrence of Environmental Consulting & Technology (ECT), the project's civil engineer and landscape architect, the low-lying southeastern corner presented the biggest challenge. Fronting the building's main entry on Iowa Avenue is a newly landscaped area, including tree beds and planter areas that act as mini bioswales. Rounding the corner of Third Street along the building's side, where a dedicated staff entrance leads into the bicycle storage and mud room, additional tree beds and rain gardens are in place — which lie beyond the building's footprint, surpassing the Water Petal's requirements for Responsible Water Use — for bioretention and to collect excess stormwater, should the building's cisterns reach capacity. "The tunnel bisects these two areas, basically right down the middle," Lawrence says.

Employing an innovative and somewhat unconventional circumvention strategy, ECT led an effort to conjoin the site's hydrology and bisected landscapes with piping that traversed the existing tunnel from below. This was done to "fortify the retention area" and maintain "a single outlet and connection point" for all bioretention at the site's lowest lying point, where the two exterior landscaping programs meet at Third and Iowa. "We opened it up," Lawrence says, referring to the landscape's capacity for storage and infiltration, which was accomplished in a rather inconspicuous manner.

This strategy left the issue of how to design the paved surfaces that line the site's public-facing sides along the avenue and street and divide the building from the exterior landscapes. The use of standard concrete or asphalt would result in excessive runoff, with only a portion of the stormwater reaching its intended destination, being subject to wind and gravity. As an

The Stanley Center's water treatment room is viewable from a small ground-floor meeting room in the building's northeast corner

alternative, the design team employed the services of concrete manufacturer Hahn Ready Mix, based in nearby Davenport, specifying a unique and durable mix design of permeable concrete that is installed beneath the brick pavers. The surface hardscape, meanwhile, is configured with small gaps between each brick to allow for easy infiltration. Not only does this enable runoff to penetrate the landscape and reach the rain gardens more efficiently, but it also acts as a stabilizing base for a critical corner of the site that is anything but level.

Lawrence admits this was the more challenging and costlier option. And yet, "we knew the technology was proven, and we felt confident that if we followed the set specifications" for installing, finishing, curing, and consolidating (the process of removing air voids that can compromise concrete's durability), "at the end of day, the product would perform as anticipated."

As both a practical and functional matter, some things just shouldn't be kept out of sight on a project like this. Positioning the cisterns above ground was the practical and economical choice considering the storm sewer's presence. It also obliged the larger water engineering team to insulate the silos in three inches of foam to prevent freezing during the winter months. In turn, the cisterns' conspicuous presence near the building's front entrance yielded dividends. "These types of industrial cisterns are not common in urban downtown settings; they're not something you would normally celebrate," says Lyndley Kent. However, they also provided the opportunity, in Kent's words, "to call attention to how this building is using water and how it's being treated. No one is trying to hide that."

In many respects, that's an understatement. The water room itself is confined but not hidden; a small corner conference room features a door-sized window that looks into the room, where all critical equipment and functions are on full display. From there, the varied but streamlined web of insulated pipes — copper pipes for water and HVAC refrigeration lines, cast iron for sanitary and storm drain lines, and steel for sprinkler lines — course their way through the building's exposed joists. For the curious, one can trace where domestic cold, domestic hot, non-potable, storm sewer, and storm sewer overflow water are being channeled throughout the building.

Also residing in plain sight is a paved ode to the creek bed. Winding and cutting its way laterally through the herringbone configuration of pavers at the front entrance is a circuitous band of dark gray pavers, five bricks wide in a running bond pattern, that is designed to mimic the path of the original creek bed. Additionally, a new stormwater overflow pipe is situated directly below the band of pavers, directing any overflow from the cisterns into the bioretention area at the low corner of the site. This creative touch makes for a great conversation piece. It is also an open invitation, extended to visitors and virtually anyone who crosses paths with the Stanley Center, to learn about the history of this site and this city, the local ecology and hydrology of the area, what has transpired over the previous two centuries that has adversely impacted those conditions, and what is presently in place to restore some fraction of what's been lost.

"These types of industrial cisterns are not common in urban downtown settings; they're not something you'd normally celebrate. [But they] call attention to how this building is using water and how it's being treated. No one's trying to hide that."

LYNDLEY KENT
Principal, Neumann Monson Architects

"That tunnel was a major challenge for the project. For a retrofit or redesign of an existing building and site, it was extremely challenging, just in a traditional sense. Now layer on the Living Building Challenge piece, and that makes it even more challenging."

JONATHAN LAWRENCE
Senior Landscape Architect, Environmental Consulting & Technology (ECT)

"If the cisterns were located underground, there would have been a big disconnect over how to connect all these systems. The storm sewer tunnel beneath the site was a barrier to this whole project and probably the most cumbersome thing we had to work around. It is right under the floor of the building, and the floor steps up and over it. It severs the building into two sides."

SALLY OBERNOLTE
Principal, Neumann Monson Architects

The Center's two 6,250-gallon cisterns stand sentry, just to the side of the building's front entrance along Iowa Ave.

"We understand the [rainfall] averages, but we also know there are fewer and fewer norms. So, we're building a system that is resilient to future change as well."

LYNDLEY KENT
Principal, Neumann Monson Architects

COLLECTION, STORAGE, AND RESPONSIBLE USE

"You have to be careful where you build in Muscatine, or you'll have more water than you want," says Warren Heidbreder, a member of Stanley Center's governance.

The Water Petal "envisions a future whereby all developments are based on the carrying capacity of the site." While the temperate stretch of the Mississippi River that is home to Muscatine receives far less rainfall than, say, Baton Rouge, average precipitation rates (up to 40 inches annually) do exceed other areas, particularly in points north. The area also averages 58 days of snow cover each year, with a humidity rate below 60 percent for half the year. These conditions, while not extraordinary or even ideal, do provide the opportunity to fulfill the Living Building Challenge's Water Petal and its Net Positive Water imperative.

The Stanley Center does indeed have the capacity to be self-sufficient in terms of water, and then some. The means and technology to harvest and store stormwater and divert overflow into bioretention areas that respect the site's unique hydrological conditions, are all in place. Of course, achieving sustainable water use patterns is about more than elegant design solutions for collecting it. How the water is channeled, treated, and eventually used by building occupants are all matters that lie at the heart of fulfilling the Water Petal's stringent standards.

Neumann Monson Architects worked closely with civil engineering and landscape architecture firm ECT, water resources engineering firm Biohabitats, and MEPT consultant Design Engineers to implement strategies that would ensure the Stanley Center operated as a holistic, net-positive water site. ECT and Biohabitats collaborated on the design of the site's rainwater harvesting system that services the building and site's potable and non-potable water needs, which includes the two 6,250-gallon cisterns as well as the Stanley Center's adjacent water room, where all critical treatment and filtration functions are housed.

"We understand the [rainfall] averages, but we also know there are fewer and fewer norms. So, we're building a system that is resilient to future change as well," says Lyndley Kent. ECT performed extensive forecast modeling to predict future events and climate-induced extremes. The latest specifications indicate that rainfall events that produce up to 1.25 inches of rain will not yield any runoff that leaves the site.

The unpredictability that comes from designing on a tight urban site such as this, where there are only so many places to redirect water, is not insignificant. "There's not a lot of space for on-site dispersal," says Shayla Woodhouse, project engineer with Biohabitats. But as a retrofit atop existing infrastructure rather than new construction, such space constraints became, according to Woodhouse's colleague, senior engineer Pete Munoz, "a big part of the sustainability story." Coordination and communication between teams and various sub-consultants were also key. Woodhouse and Munoz, whose company has

"There is not a lot of existing space to do any sort of on-site dispersal. So, with the constrained space we looked at an evaporative greenhouse as a way to discharge the wastewater through evaporation. This was going to be in the center of the building, but it was value engineered out due to costs."

SHAYLA WOODHOUSE
Project Engineer, Biohabitats

consulted on nearly a dozen Living Building Challenge projects, describe it as "seamless." "We spent a lot of time understanding and coordinating how water moves between systems on this site," Munoz says, such as where to place drainage points on the roof and how to stagger and locate filtration systems. "It's those little things that can make or break a project. Integration is always important for Living Buildings."

Rainwater runoff from the rooftop and mounted solar panels — including the panels that comprise both canopies along the front and side — is pre-treated via an 800-micron mesh screen that catches leaves and other detritus, preventing unwanted particulates from entering the cistern. (Particles larger than 800 microns are flushed into the storm drain.) The filtered rainwater is then conveyed to above-ground cisterns for storage. As water is needed in the building, it is pumped through an advanced treatment system to create safe potable water.

This multi-barrier treatment system, supplied by Ecovie Water Management, comprises a calibrated sequence of filtration processes. Harvested water first travels through a self-cleaning sediment filter with a 30-micron mesh screen, followed by a 20-micron carbon block filter to remove bad taste and odors, a 5-micron polypropylene turbidity filter to remove small particles, and finally a charged-membrane filter to provide microbial reduction and aid in microbiological log reduction value (LRV), which is a measure of how thoroughly the decontamination process reduces the concentration of contaminants. All this is followed by UV disinfection to prevent waterborne disease, and allow for pH adjustment and an option for secondary disinfection. Lastly, the water is held in pressure tanks ready for distribution to meet potable and non-potable demands.

Overall, and wherever possible, most of the water that finds its way onto the Stanley Center site is collected, stored, and treated until it is needed. This highly efficient set of systems

yields the added benefit of relieving a significant portion of the Stanley Center's dependence (and impact) on municipal infrastructure. (It should be noted that the Stanley Center building remains connected to the city's municipal wastewater system.) With the cisterns, permeable hardscape, and bioswales in place, the project will, on an annual basis, retain more than 200,000 gallons on-site through infiltration and evaporation.

The Stanley Center's conservation efforts also extend to the investment the organization has made in its neighbor, the Muscatine Center for Social Action (MCSA). Through the Center's "handprinting" efforts (covered in detail in the next chapter), which measurably lessens MCSA's overall water usage thanks to new high-efficiency fixtures, the likelihood of combined sewer overflow (CSO) events is reduced, which in turn reduces the overall volume that the municipal stormwater system has to account for.

> *"This is where constraints become opportunities. We really had to adjust because of budget reasons. The pragmatic side informed the location of things like restrooms, even though the original intent may have been different."*
>
> **KHALID KHAN**
> Principal, Neumann Monson Architects

WATER DIAGRAM

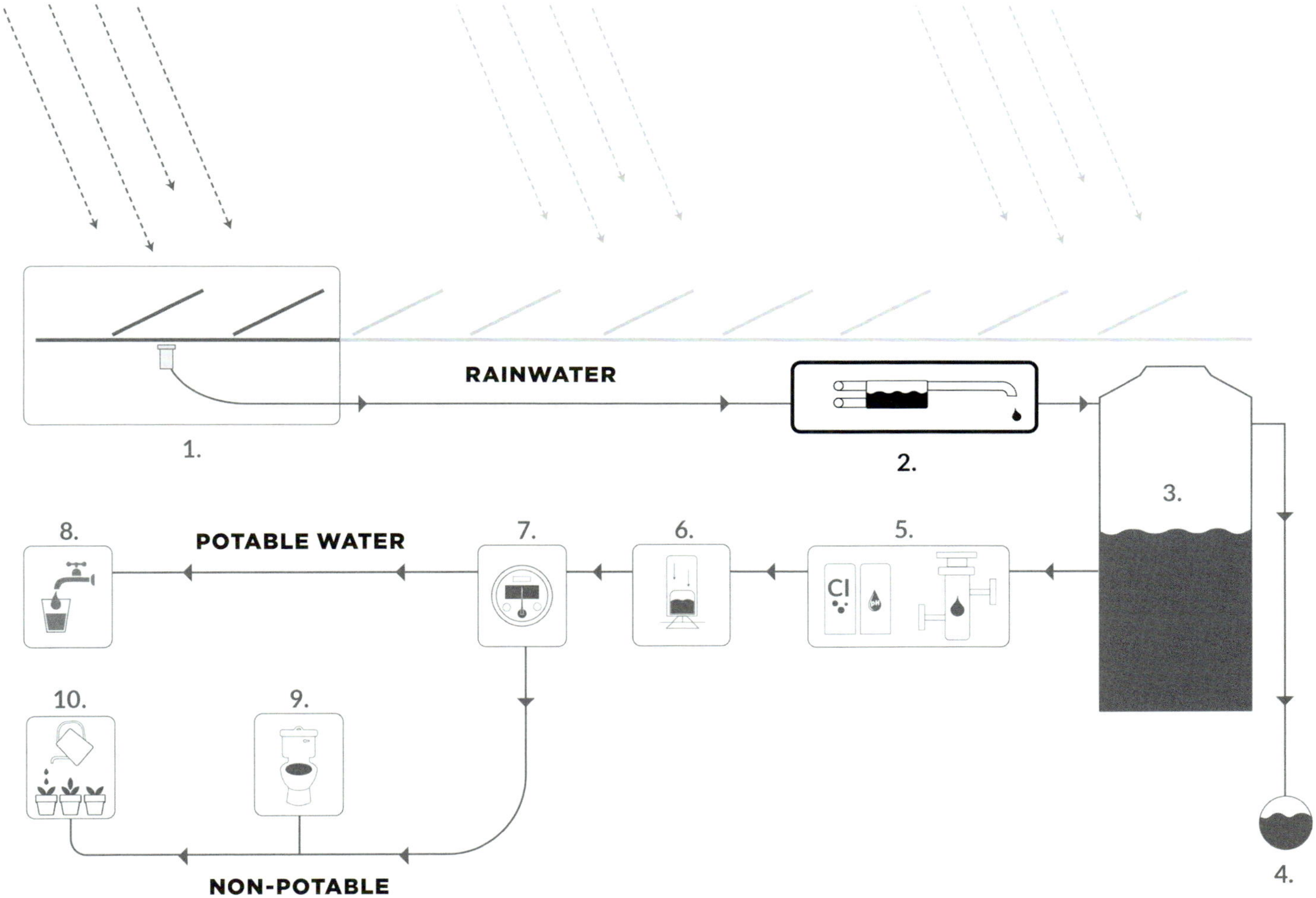

1. Rainwater from PV (Solar) Array and Roof
2. **Hydraulic Jump Filter**
3. Rain Collection in Cisterns
4. Overflow to On-Site Stormwater System and Bioswale
5. Multi-barrier Filtration with Disinfection and pH Monitoring
6. Pressure Tank
7. Meters
8. Indoor Potable (Drinkable) Water
9. Non-potable Plumbing Fixtures
10. Non-potable Irrigation

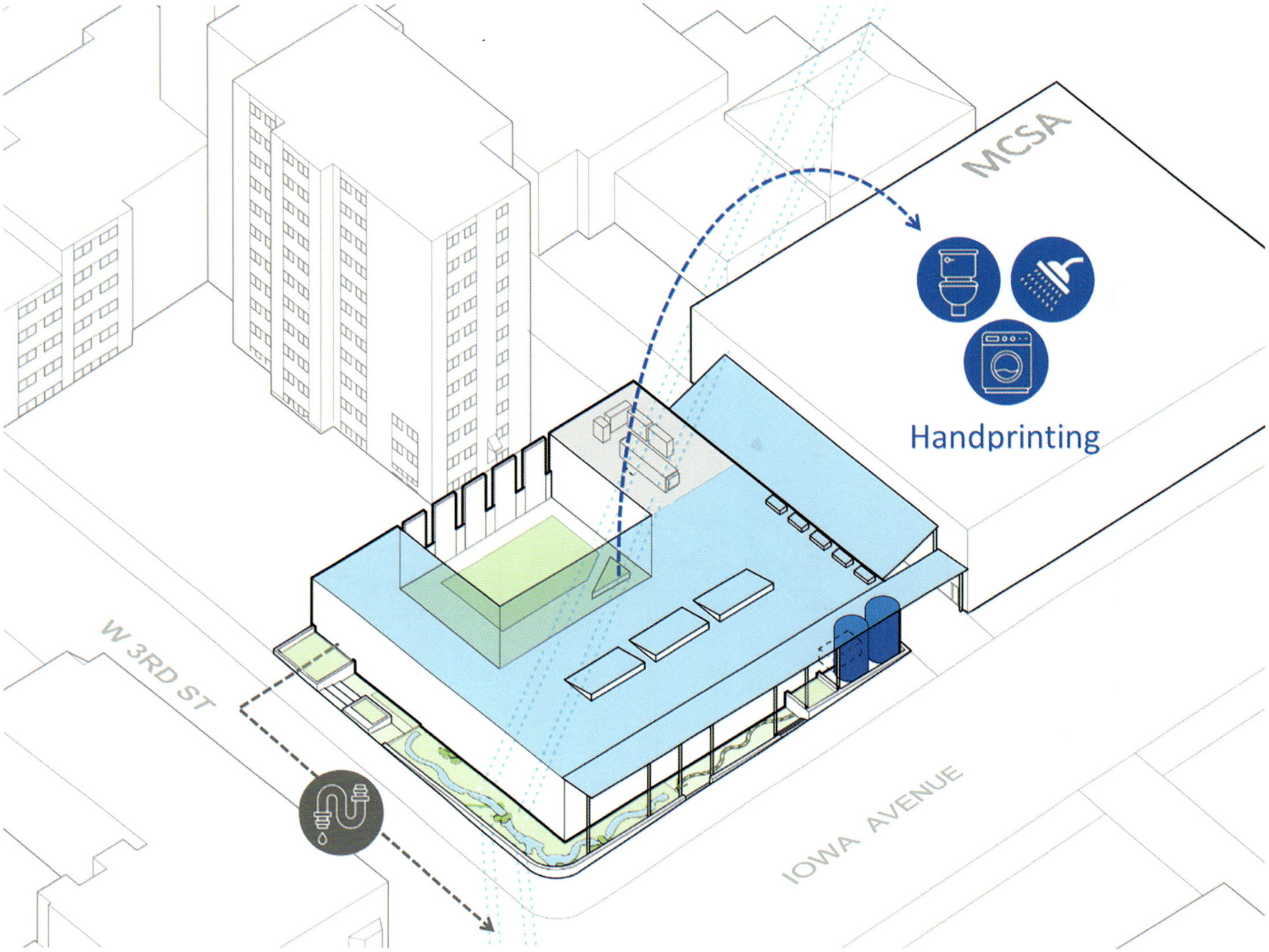

New water-efficient fixtures being delivered to MCSA, as part of Stanley Center's "handprinting" partnership with its neighbor

"The Stanley Center is our partner in the water usage, or should I say reduced water usage. They came in and financially helped MCSA with some of our plumbing needs, especially back in the shelter area. So, it reduces the amount of water that is used in our building."

ROBERT ALBEE
Interim Director, Muscatine Center for Social Action (MCSA)

"Beginning from the schematic design phase... we had a lot of different options on the table for servicing the building's water needs and meeting the Petal's Responsible Water Use Imperative, and two of those options got us to full [Living Building] certification. One was handprinting and the other was the on-site greenhouse. So, it's always a process to go through."

LYNDLEY KENT
Principal, Neumann Monson Architects

HANDPRINTING WITH MCSA

Like the project itself and the client leading the way, Neumann Monson Architects had grand ambitions. And practically speaking, to everyone's credit, most of them came to fruition.

A few others did not. "We had extensive plans for a greenhouse and constructed wetlands that would serve as [on-site] wastewater treatment," says Lyndley Kent. The greenhouse, originally proposed to be situated in the building's center, would have been an effective measure for discharging wastewater through evaporation. (It also would have likely housed the Stanley Center's urban agriculture component and precluded the need for the courtyard.) Once the value engineering process began, however, such ideas proved cost prohibitive and fell by the wayside. "That's when the idea of handprinting started."

In the context of the Living Building Challenge, a "handprint" measures the positives that a building produces or causes through standard operations, relative to business as usual. Opposed to a footprint (as in one's "carbon footprint"), which considers the negative impacts of building and manufacturing something, a handprint considers the impacts of generating net positives, with one considerable caveat. Those positives must extend beyond the scope of the building and its owner's on-site operations.

The Stanley Center has become a "handprinting" organization, which means that its sustainability efforts are measured both in the technologies applied to harvest, treat, conserve, and divert water (environmental action), and in the genuine good the organization commits to performing within its community (social action). For the latter, the Stanley Center committed to upgrading outdated water fixtures in the neighboring MCSA building.

According to Kent and her colleagues at Neumann Monson, handprinting was the better option to pursue, both economically and philosophically. It aligned with the Stanley Center's values and stated commitment to give back to Muscatine. While a more robust, sophisticated, and costly wastewater treatment system would have further relieved the building's burden on municipal infrastructure, handprinting has immediate impacts for the patrons and staff of MCSA, an organization that provides temporary housing, basic health care, and educational and support services to those in need in Muscatine County. Consequently, for a facility with significant water demands, installing new and efficient showerheads, toilets, and washers will ease some of MCSA's financial burden as well.

As mentioned in the previous chapter, the Stanley Center is connected to Muscatine's municipal wastewater system, which by most accounts is a relatively clean and efficient one. Even with design measures in place that divert significant quantities of stormwater from the city's storm sewer, extreme precipitation events are inevitable, as is some degree of reliance on the city's waste management infrastructure. All the same, when there's runoff, there are negative impacts. And wherever and however sanitary wastewater is treated off-site, there are said impacts. Those and related impacts are used, in part, to measure a building's water footprint. The Stanley Center is offsetting that footprint with its handprint. Greater water efficiency throughout MCSA helps reduce the building's potable water demand as well as the wastewater that is sent into Muscatine's wastewater system.

Upgrading water fixtures throughout MCSA's building relieves much of the organization's demand and associated utility costs. And over time, this conservation measure neutralizes the impacts of what is sent into the municipal system. Everyone benefits.

THE ENERGY PETAL

A Roof That Isn't Big Enough

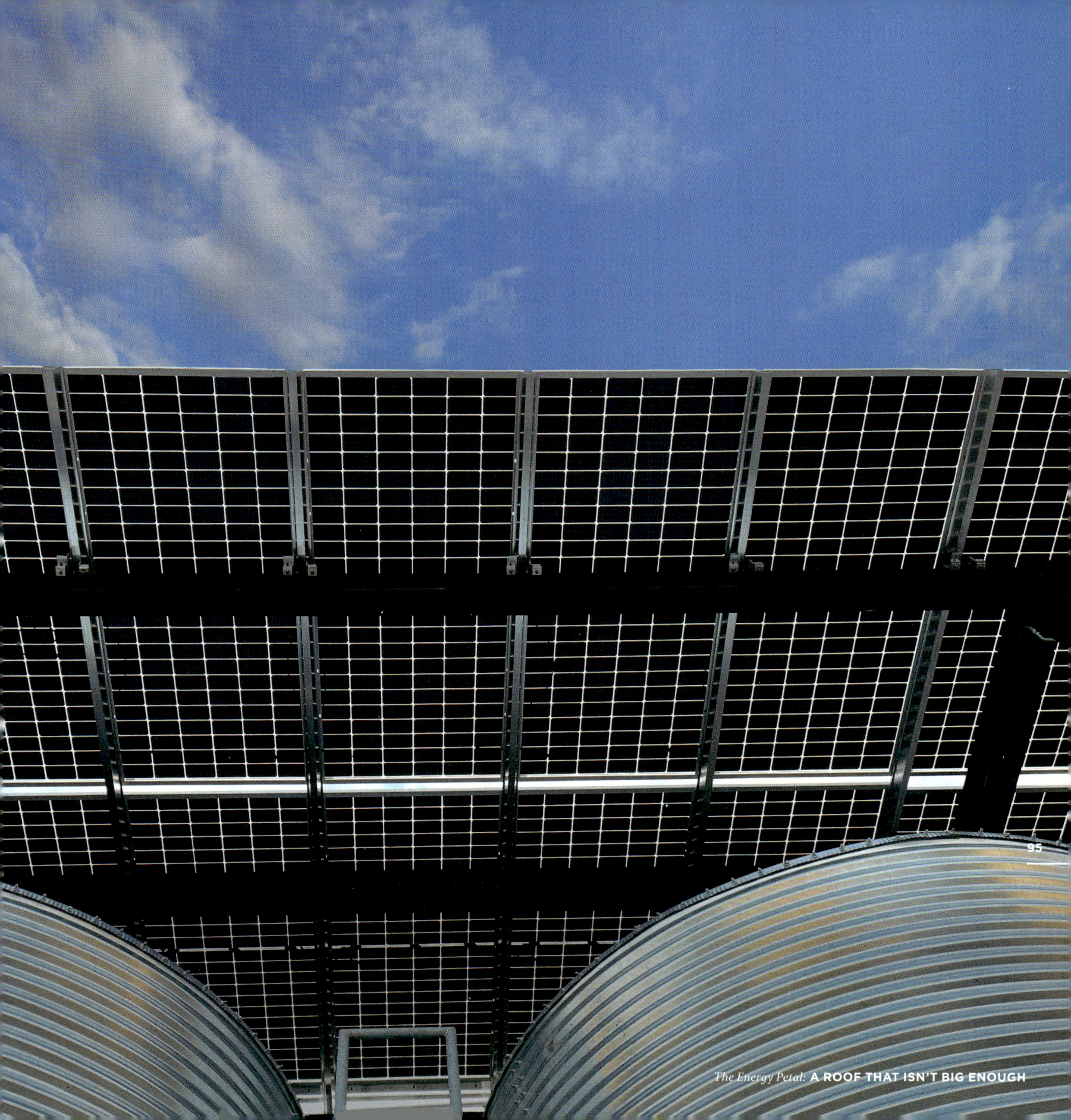

Stanley Center
FOR PEACE AND SECURITY

"The roof wasn't large enough to handle all the solar needs. The end result was the two canopies, but earlier on we did a brief study and looked at scale jumping, which would have put some of the solar on the roof of MCSA next door. We looked at the pros and cons of that option, but quickly turned back to the idea of canopies and making the system more visible."

SALLY OBERNOLTE
Principal, Neumann Monson Architects

A ROOF THAT ISN'T BIG ENOUGH

The rooftop of the Stanley Center's new building wasn't a problem, strictly speaking, but it also didn't present the design team with a lot of flexibility in terms of solving the building's renewable energy demands, not to mention stormwater collection and the stringent material requirements for achieving a Living Building.

Any roofing solutions had to accommodate several interrelated compliance demands. And such solutions had to meet a tight schedule to ensure construction timelines for both the building's interior and exterior progressed in tandem.

"The site itself is limited. The building takes up most of the available property and making this building net-positive energy and with on-site generated energy, that's very difficult," says Tim Lentz, a mechanical engineer with Cedar Rapids-based Design Engineers, the project's consulting MEPT engineer. "You need about as much roof square footage as you have occupied square footage in the building." Judging by existing conditions, this math wouldn't add up.

It all came down to feasibility. How large of a solar array could realistically be mounted to the Stanley Center's rooftop and generate enough wattage to surpass the building's energy demands, all while leaving enough room for requisite mechanical equipment? Further complicating matters were the introduction of the 2,469-square-foot courtyard (which

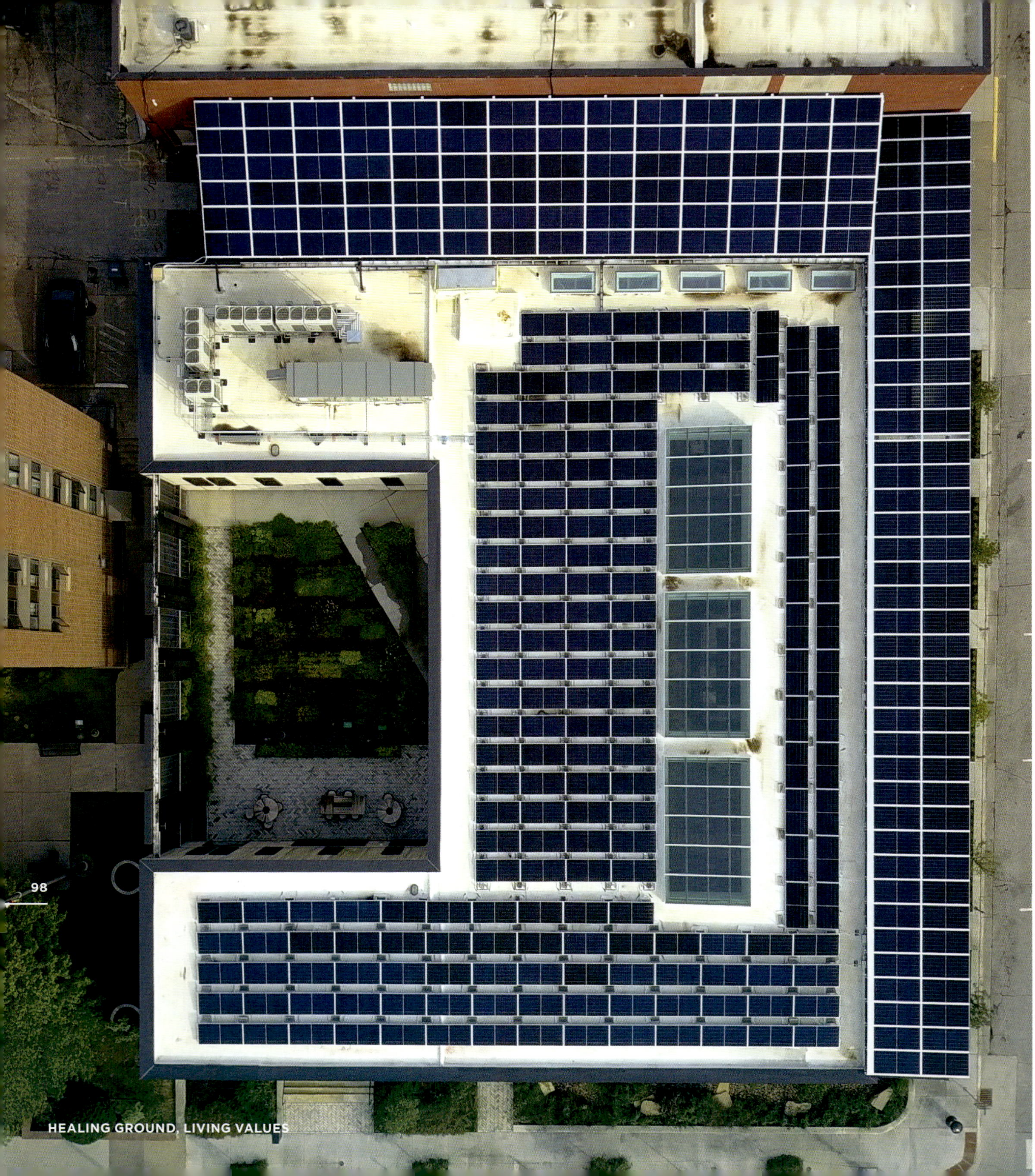

"The site itself is limited. The building takes up most of the available property and making this building net-positive energy and with on-site generated energy, that's very difficult. You need about as much roof square footage as you have occupied square footage in the building."

TIM LENTZ
Senior Sustainability Engineer, Design Engineers

"The solar panels are far more visible with the canopies. Had we gone with a different option, they wouldn't have been visible at all from the street. They would have all been sitting on the roofs of both Stanley Center and the MCSA, and you wouldn't know they were there. This way, with the canopies, they become a teaching tool."

SALLY OBERNOLTE
Principal, Neumann Monson Architects

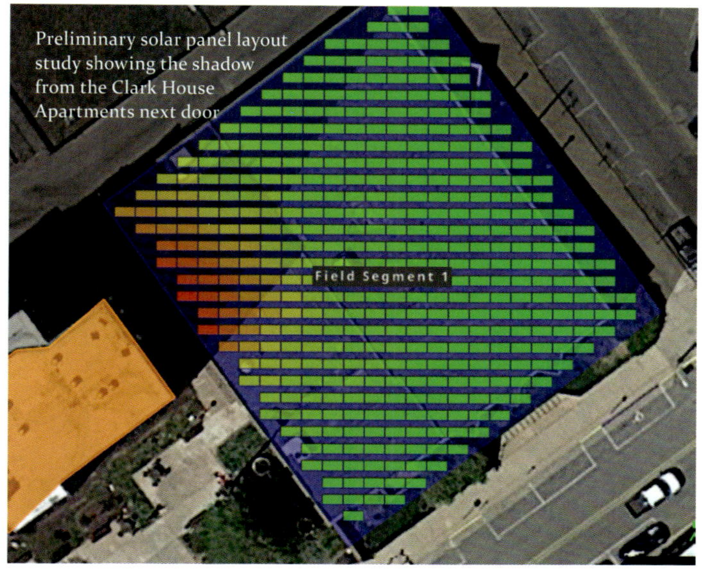

Preliminary solar panel layout study showing the shadow from the Clark House Apartments next door

removed approximately one-fourth of the existing building's available surface area) and the installation of two perpendicular rows of skylights at points where sun exposure is a premium, especially in the warmer months. In short, there simply was not enough space.

The design team at Neumann Monson concedes that off-site options for renewable energy production were a topic of discussion early on. While off-site renewables are a viable option within the Energy Petal, the nature of maintaining off-site equipment and transmitting that generated energy directly to the project (and nowhere else) can be cumbersome. So, back to the roof.

During those early design phases, the roof question evolved from a logistical concern into a more holistic discussion about placemaking and equitable development. Two sets of panels were specified: a ballast-mounted panel system that sits shallow to the roof, with individual panels angled for maximum exposure; and the awning system suspended over the building's northern and eastern edges, comprising bi-facial modules that collect direct sunlight and reflected sunlight from below. (The resulting energy boost from reflected sunlight, while nominal, occurs with greater intensity along the building's front entrance when compared to the awning over the alleyway.) The ballasted system is carefully designed around the skylights and the rooftop's far western corner, where mechanical equipment resides in a heavily shaded pocket, owing to shadow cast by the adjacent, 11-story Clark House. With each of these systems working together, the projected on-site solar energy production of 186,000 kWh annually is designed to provide 108 percent of the Center's energy needs.

Supporting such a massive solar array (332 total panels, with a capacity of 149.4 kW) also required an especially sturdy and Red List-compliant roofing material. The roof assembly specification features a non-halogenated, rigid foam (polyisocyanurate) insulation material and membrane system — manufactured by GAF and installed by Black Hawk Roof Company — that together provide a durable substrate for the solar array. The roof membrane is critical for the collection of water for potable use. This composite roof system contains no hazardous flame-retardant chemicals, holds a Health Product Declaration (HPD), and is NSF P151 compliant, thus ensuring that rainwater catchment products will not impart contaminants into the potable water supply that exceed EPA regulations for drinking water. Prior to the installation of the solar array, the roof was layered with 100 4x4 squares of durable coated glass fiber facers (CGF), also known as cover boards. These materials are lightweight, have high compressive strength, excellent dimensional stability, and are highly resistant to mold.

Of course, even with a stable rooftop in place that is newly equipped for renewable energy production and a steady influx of natural light for the building's interior, ensuring all that solar energy is properly transmitted and stored for later use presented a new set of mechanical, technical, and design challenges.

"This [energy] recovery system is by far the most efficient one I've ever specified. It helps make the building 97 percent efficient, whereas most buildings are in the 70th percentile range."

TIM LENTZ
Senior Sustainability Engineer, Design Engineers

"The solar array has module level control and strategic wiring. Shading on one portion of the array isn't going to reduce the output on the entire system. The inverters tie into the batteries and the grid, they charge and manage the batteries and deploy them to emergency loads when necessary. It's all internal; the inverter makes the decision of where to send the energy."

TIM LENTZ
Senior Sustainability Engineer, Design Engineers

SAVINGS AND RECOVERY

There are a handful of large, investor-owned public utility companies in Iowa, providing natural gas, electricity, and transmission services to most of the state. Iowa's renewable energy industry is robust, thanks in large part to the growing number of wind farms and turbines, which according to recent data have supplied Iowa with approximately 60 percent of all electricity generated in the state. The Stanley Center, however, remains a strictly local operation.

While entirely self-sufficient from an annual perspective, the building's on-site grid is connected to municipal utility provider Muscatine Power and Water (MPW) for the purposes of delivering excess power back to the local grid and drawing power after sundown or on days with insufficient sunlight. MPW is a relatively small provider, but a forward-thinking one nonetheless. Its Choose Green Muscatine program enables customers to invest in wind power via renewable energy certificates (RECs). And now, with this new Stanley Center tapped in and boasting a highly visible solar array, positive and renewable change on a much larger scale feels imminent.

All big changes start with a ripple. This effect came in the form of MPW needing to create a new rating class for large commercial solar arrays. Previously, net metering rules set by the utility were geared only to smaller arrays, typically tied to single family homes. Design Engineers and the Stanley Center collaborated with the utility provider to develop a new commercial array class, which has the potential to sway other commercial facilities to act in kind by building and hard wiring their own arrays, whether on-site or at an off-site solar garden or similar scheme. And while generating more electricity than is required for a building and returning that excess power to the grid may not be necessary for all buildings, inspiring more building operators and owners to wean themselves from any power source that burns fossil fuels is undoubtedly one of the desired outcomes of the Living Building Challenge.

To maximize solar power generation, a large portion of the Stanley Center's vast array is fitted with micro inverters, which include a maximum power point tracking system. As opposed to a string inverter system, in which the collective modules are limited by the production of single panels, which can be hampered by shading and other external factors, a micro inverter system controls each module individually. This means that the entire system's production capacity never becomes limited by any low-performing modules, due to shading or otherwise. String inverters, with their own maximum power point trackers, are used elsewhere in the array where shading is less likely to occur. "It's a lowest common denominator system, so we are trying to increase the amount of module-level control," says Tim Lentz. "One portion of the system being shaded isn't going to bring down the entire system."

The string inverters tie into the batteries. They charge and manage the batteries and deploy them to emergency loads. "It's all internal. The inverter makes the decision," Lentz says.

Lentz recognizes that while this kind of system isn't entirely new, it is relatively new within the market and makes for a far more efficient (and resilient) mode of power generation, recovery, and storage. "Battery technology itself is ever evolving and ever changing," he says. In the case of the Stanley Center, a lithium-iron-phosphate battery system was specified. Such systems are much heavier than standard lithium-ion systems, but "are ideal" for utility scale stationary applications and

To compensate for lack of square footage, owing to the new courtyard, new skylights, and rooftop mechanical equipment, dual canopies composed of bi-facial solar modules now extend over the alleyway between MCSA and Stanley Center and the building's front entrance along Iowa Ave.

ENERGY DIAGRAM

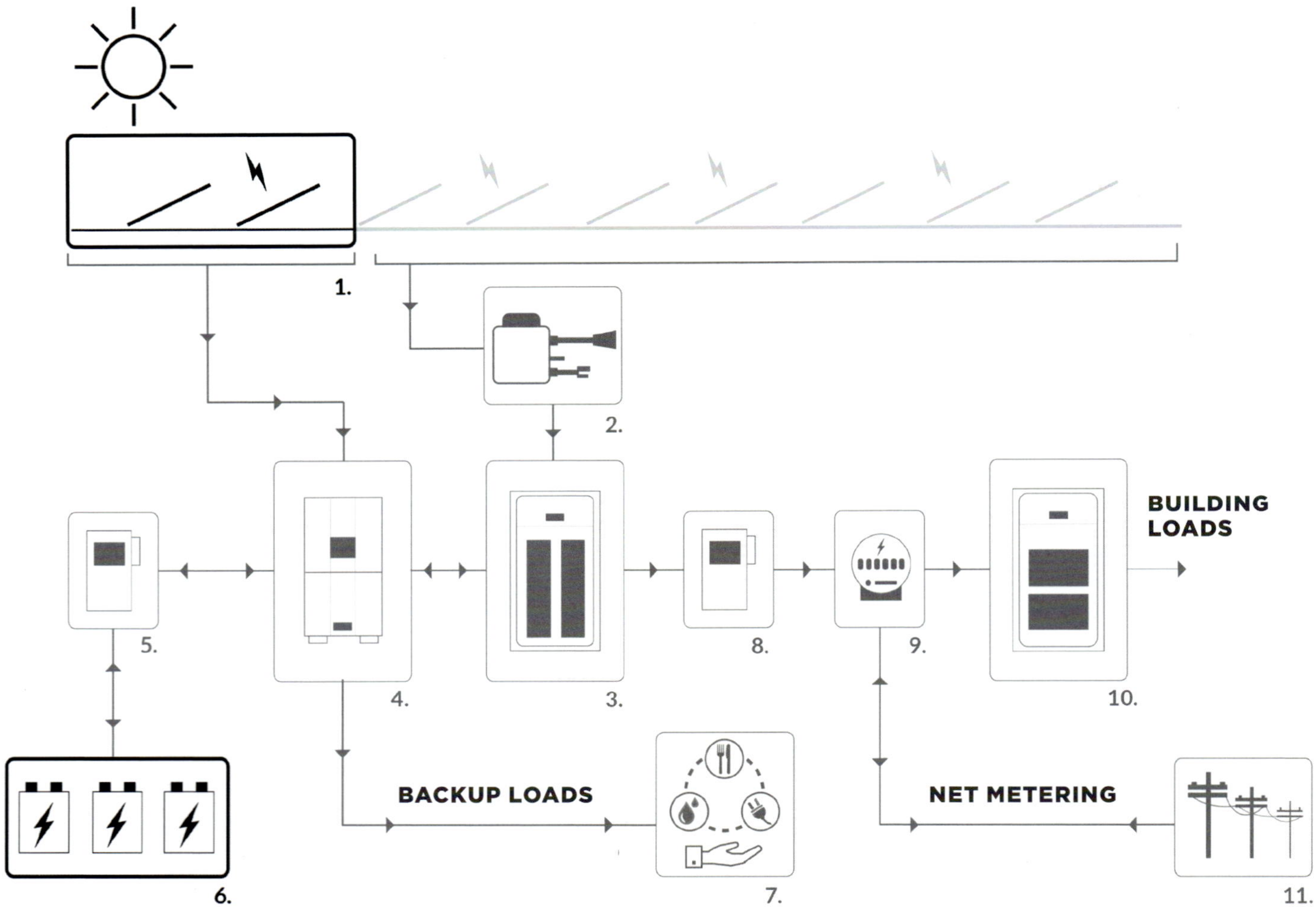

1. **Photovoltaic Array (Solar Panels)**
2. Microinverters
3. AC Combiner Panel
4. Dual Function Inverter
5. DC Disconnect
6. **Battery Bank**
7. Habitable Zone
8. AC Disconnect
9. Electrical Utility Meter
10. Main Distribution Panel
11. City Power Grid

Interior of the Stanley Center's battery storage room

backup power systems. Lithium-iron-phosphate batteries also offer longer lifecycles, low toxicity, and favorable safety ratings.

To minimize HVAC energy needed from the solar array, the Stanley Center's HVAC system consists of a high-efficiency variable refrigerant flow (VRF) system, which performs similarly to ductless mini-splits but is centralized rather than zoned, and a relatively new regenerative energy recovery system for the building's ventilation air. Many buildings have an energy recovery unit (ERU) for ventilation, which is an air-to-air heat exchanger that transfers sensible and latent heat between the outdoor air coming into the building and the exhaust air leaving the building. In the case of the Stanley Center, "this new type of ERU is by far the most efficient one I've ever specified," Lentz says. "It is 97 percent efficient, whereas most ERUs are in the 70th percentile range." This type of ERU achieves such a high efficiency with a novel regenerative heat exchanger component in addition to a common energy recovery wheel component. The regenerative heat exchanger component consists of fast-acting dampers that briefly reverse incoming and outgoing air streams every minute to regenerate energy recovery capacity without allowing stale air to recirculate in the building. This technology also prevents incoming air — especially in colder months — from interfering with the system's ability to provide neutral temperature ventilation air.

In climate zones that are subject to frost like Climate Zone 5, where Muscatine resides (according to the International Code Council's definition of zones, Zone 5 includes southern and eastern Iowa), energy recovery ventilation equipment typically requires direct electric pre-heating of incoming air to prevent frost formation inside the equipment. This preheating process is expensive and energy intensive, and an alternate frost protection method of slowing down the wheel rotation would reduce the discharge air temperature too much to be compatible with VRF equipment. But at the Stanley Center, the need for any pre-heating was eliminated thanks to the regenerative heat exchanger component of the ERU, allowing for ideal interior thermal conditions, no slowdown of the equipment, no risk of frost on the wheel, the absolute minimum energy demand, and full operational capacity regardless of outside conditions.

"Concrete is the most detrimental to embodied carbon, which is almost an understatement. We needed to make sure that we were controlling the amounts and types of concrete that needed to be added to the building."

BRANDON FETTES
Intern Architect, Neumann Monson Architects

DESIGNING A PASSIVE SYSTEM

Passive design strategies exist to provide buildings of any kind with high efficiency thermal control, air control, radiation control, and moisture control.

Broken down accordingly, this translates into a building having a high-performance enclosure for good insulation (thermal); an airtight envelope and continuous, balanced ventilation (air); high-performance windows with a balance of daylighting and shading to improve indoor environmental quality (radiation); and the mitigation of excess vapor and humidity (moisture). Working together, and when implemented correctly, these principles ensure occupant comfort and health all while keeping the building's energy costs extremely low relative to its square footage, daily usage, and climate zone.

The predicted Energy Use Intensity (pEUI) of the Stanley Center is 30.7 kBTU per square foot per year, which is incredibly low and efficient for a commercial office space. (kBTU is a unit of energy equivalent to 1,000 BTUs.) For comparable reference, EPA Energy Star tracks the median EUI for all building types in the United States, and the average office building requires 52.9 kBTU/ft2/yr. (Only warehouses, various storage facilities and distribution centers, and places of worship perform at rates relatively comparable to the Stanley Center.)

So, how was this level of efficiency achieved? The subject of building materials was obviously front and center throughout the design process, from enclosures and glazing to insulation and plumbing fixtures. But within each of these conversations and material analyses, the subtext (and, on occasion, the topic at hand) was always embodied carbon.

For the building's exterior, Neumann Monson conducted a careful materials comparison to get a clearer sense of the overall cost, maintenance needs, biophilic benefits, and embodied energy associated with various materials within the context of the building site and surrounding neighborhood. Wood, metal, glass, and stone were on the short list. (Concrete, owing to the material's high emissions associated with manufacturing, particularly when using portland cement, did not make the list.)

The potential benefits of using stone — natural material, contextual, and low embodied footprint — were collectively trumped by its exorbitant cost. Glass and metal, while both relatively low maintenance (metal more so), both come with a high footprint and offer little biophilic appeal. Lastly, wood checked off the right boxes: biophilia, low embodied energy, and low cost (up to $50 per square foot, excluding any rain screen framing and insulation). Only maintenance and local context were possible demerits associated with the design team's use of wood cladding. However, the use of Accoya siding, a specialized, non-toxic product engineered to mitigate any longevity and maintenance concerns typically tied to traditional wood cladding, helped alleviate this problem. Further, the innovative and time-honored technique of Shou Sugi Ban (aka Yakisugi), the traditional Japanese process of charring the wood, was used to add an extra layer of outside protection. It also lends the Stanley Center's exterior a striking black/gray-toned building skin that fits in well within Muscatine's industrial context.

Of course, any consideration of embodied carbon and energy must focus on what's being added to the existing building and what components are being reused. "Concrete is the most detrimental to embodied carbon, which is almost an understatement," says Brandon Fettes, an intern architect with Neumann Monson who was instrumental in conducting the building's lifecycle assessment (LCA). "We needed to make sure that we were controlling the amounts and types of concrete that needed to be added to the building."

The result of the LCA confirmed that nearly 94 percent of the building by mass was reused, which led to saving approximately two-thirds the amount of embodied carbon that a typical new construction building of similar size would have consumed (+854,037 Kg/CO_2 for a new building versus +193,545 Kg/CO_2 for the Stanley Center, not counting sequestration values).

Insulation was another matter of paramount concern. In the building's previous form, the Musser Library included only three-quarters of an inch of insulation material; the retrofitted building now features a continuous GreenGirt rain screen system with five inches of Rockwool insulation. Compared to more traditional insulation material, like fiberglass, Rockwool is largely composed of recycled materials and is highly durable, offers better heat retention, acoustic comfort, and resistance to mold and mildew. This system also added considerable bulk to the structure's final form, which in turn made for deeper pockets for the rows of punched windows, which is particularly noticeable along the Third Street-facing façade. This substantial profile creates much-needed shading and thermal comfort for employees who occupy offices in the southerly portions of the building.

The eastern and northern faces of the building are a different story entirely. The solar canopy along Iowa Avenue provides a visually striking overhang, as well as an important shading device for employees and visitors using the front entrance. Whatever amount of natural light doesn't enter the building because of this canopy is more than offset by the new skylights suspended over the Stanley Center's double-height atrium and "living room" area. But along the side of the building that abuts the alleyway, where more private offices line the perimeter, natural light is hard to come by. Or at least it was.

With the alley's new solar canopy in place, punching a series of new windows into the building wasn't a realistic option, to say nothing of the fact that offices looking out into an alley is less than ideal. The solution was more skylights. Within each office along the northwest face, a rectangular skylight looms overhead, bringing daylight into the enclosed spaces. As a bonus, these offices have the distinction of featuring their own irrigated green walls, each one suspended under its respective skylight. This simple yet innovative design intervention adds biophilia, thermal comfort, and a wellness focus to a part of the building's interior that might otherwise be undesirable.

FIRST FLOOR PLAN

1. Solar Battery Storage
2. Water Treatment
3. Habitable Zone
4. Flex Space During Emergency Use
5. Restroom/Shower - Water Access
6. Lockers
7. Food/Supply Storage

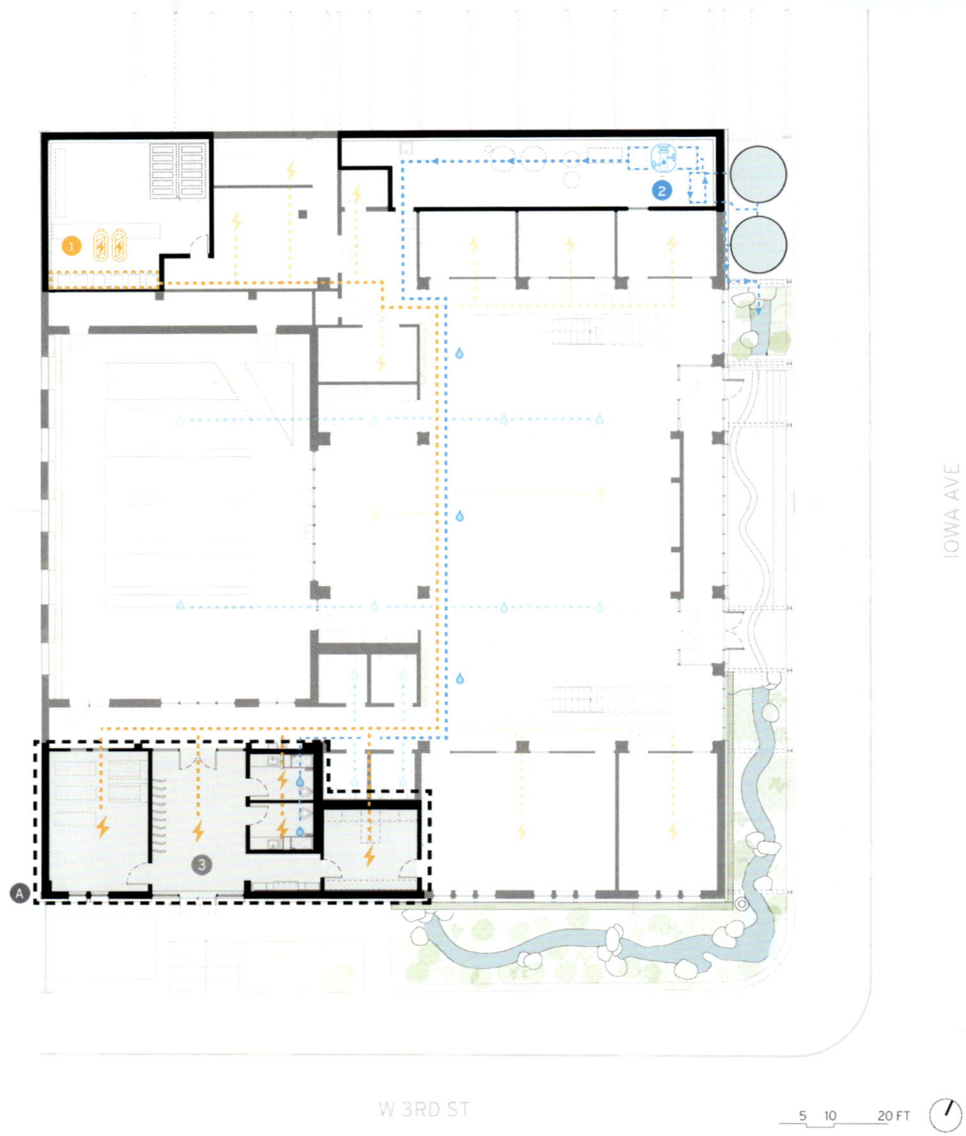

ENLARGED PLAN FIRST FLOOR HABITABLE ZONE

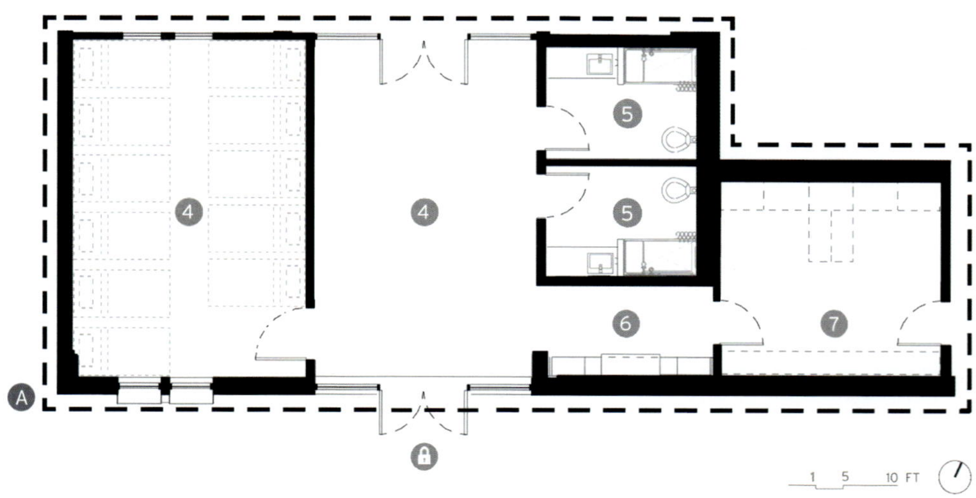

"We looked closely at our resiliency strategy and knew it was unlikely that all employees would be in the building during an emergency. The Stanley Center can be a refuge. It was designed to meet those basic human needs in the event of a natural disaster or a long-term power loss. People can have a safe space for power, water, and shelter."

LYNDLEY KENT
Principal, Neumann Monson Architects

ENERGY PETAL OVERVIEW

- 149.4 kW photovoltaic solar array
- 332 solar panels
- 186,000 kWh generated annually

RESILIENCY STRATEGY

A passive building that is cool in the summer and warm in the winter, all without placing much strain on mechanical systems, has purpose that runs deeper than ensuring occupant comfort. The Stanley Center is both durable and high performing, it is energy and water efficient, it is carbon positive, and it is self-sustaining.

Owing to all this and more, in the event of an emergency, the facility is equipped to function as a bastion of sustainability, providing building occupants with power, access to heat, drinking water, shelter, and other critical resources.

As sustainable as the building is, practically speaking it was not feasible to back up the entire facility (comprising power, water, heating, cooling, and food, all operating at full tilt) should such an emergency occur, and the Stanley Center becomes obliged to shelter dozens of people. While the building remains connected to the municipal grid for the purposes of net metering and nighttime power usage, should the grid fail, a small backup power bank resides in the building's battery room, which can support a small portion of the building — three rooms that comprise the designated emergency shelter — with electricity, drinking water, and working restrooms.

The three rooms in question occupy the area of the building immediately accessible via the employee entrance, discreetly situated on the side of the Stanley Center facing Third Street. From there, the building's mud room, fitness room, coat and locker room, and two adjacent bathroom facilities with showers comprise the primary shelter area. And thanks to the building's two 6,250-gallon cisterns, water usage in such a scenario — while rationed appropriately — would not be a grave concern. Either way, in a pinch, the Stanley Center could well function as a safe haven for Stanley Center employees in need.

In the event of a natural disaster, the building and its systems were designed to meet the needs of full-time occupants at 75 percent capacity. "We looked closely at our resiliency strategy and knew that it was unlikely that all employees would be in the building during an emergency," says Neumann Monson's Lyndley Kent. "The Stanley Center can be a refuge. It was designed to meet those basic human needs and more in the event of a natural disaster and a loss of power long term. Occupants can have a safe space for power, water, and shelter."

THE HEALTH + HAPPINESS PETAL

Discovering Purpose

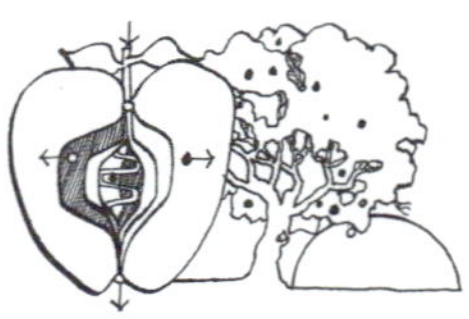

EXIT

Staff from Stanley Center, Neumann Monson, and Graham Construction gather to fill custom shadow boxes with pearl button blanks for the building's kitchen backsplashes

DISCOVERING PURPOSE

The ways in which we experience a building and designed spaces, whether through the lens of a resident, employee, or visitor, have become integral facets to how design professionals market their entire philosophy.

In so many words, a building should by design have the capacity to inspire and motivate. A building should feel approachable from the outside and intuitive on the inside, carefully blending components that recognize the beneficial impacts of human scaled design. It must ensure good health through optimal indoor environmental quality, good circadian rhythms through access to natural light, and whole-person wellness by using the principles of biophilic design.

Of course, all of the above comprises the outlier relative to the building industry writ large. Historically, at least within the modern era, buildings have largely been treated as functional cogs of the social and political wheel. The physical and psychological well-being of a building's inhabitants have not been prioritized. In many instances, this remains the case today. But it is not a universal condition, and the Living Building Challenge is proof of that. Considerations like fair and equitable labor practices, questions about how and from where building materials are sourced, and whether said practices and materials contribute to better building performance and circular economies are all key pieces to the puzzle.

In time, building "green" became synonymous with building healthy. It denotes building for a long life, while prioritizing interior materials and finishes with little to no volatile organic compounds (VOCs), and accounting for end-of-life pathways for emissions-intensive building materials. It means looking for opportunities to innovate and introduce carbon-sequestering, cradle-to-cradle solutions into buildings, as opposed to standard cradle-to-grave practices, which result in larger landfills and the release of embodied carbon and methane.

Arriving at this mode of thinking has required a gradual yet tectonic shift in thinking on the part of the design and building community to adopt more holistic practices. Nearly two decades ago, upon founding the non-profit Architecture 2030 (developer of the 2030 Challenge), Edward Mazria said, "I realized that we have the solutions, but nobody had put them all together."

The pieces are still coming together to this day. After all, this particular Petal of the Living Building Challenge specifies health and happiness. Relative to where the building industry was just a few decades ago, this industry is light years ahead of the previous generation in terms of understanding the health impacts — both direct and indirect — of practically every building material on the market.

These advancements have largely been aided by the use of Environmental Product Declarations (EPDs), which are documents that quantify environmental and lifecycle impact data — including embodied carbon — from manufacturing a product. The evolved version of this trend has come in the form of the International Living Future Institute's Declare label, aka a "nutrition label," which is a voluntary disclosure tool for manufacturers to publicly share the ingredients of building materials. While all seven Petals are fused and interdependent, a convincing case could be made that Health + Happiness and Materials have more mutual overlap than most other pairings.

Happiness, on the other hand, is more abstract. Health is a target you can hit using tangible and proven design strategies such as improving indoor air quality; Happiness is a feeling or sensation you want to facilitate using an unspecified range of strategies.

But what is happiness, after all, if not the product of good health? And how does one design a healthy and wellness-focused environment without introducing access to daylight and connections to nature and natural elements? Achieving the Health + Happiness Petal is very much about fulfilling basic human needs and embracing nature. It has just taken a long time for the building industry to make the connection between what constitutes a green building and a healthy building. As it turns out, they're not mutually exclusive.

"This wasn't just about being neutral or being less bad. This is about creating a building that gives back. A building with a net-positive footprint, a building that collects all the water that falls on-site, a building that in demonstrable ways makes its neighborhoods better. And a building that contributes to the health and happiness of the people that work in it. I mean, all of that just resonated so much with us. At first we thought, is that a thing? Is this really something that people do? It was just amazing."

KEITH PORTER
CEO and President, Stanley Center for Peace and Security

DESIGNING A LIVING ROOM

The Stanley Center is something of an anomaly. The organization's new home at Third and Iowa is a private office building, first and foremost, but with a striking façade and human-scaled interior layout that beckons visitors to stop and ponder this highly engaging built environment.

Its front entrance is elegant and modest, leading one into a voluminous double-height atrium that literally bursts with greenery. Open concept spaces give way to small nodes and a variety of seating areas. This office building's front desk does not exist, having been replaced, in a sense, by a curated mix of banquettes, bar stools, conference tables, and interconnected living rooms.

Quite fittingly, the Stanley Center's entire foyer feels like a well thought out community space one might find in a new public library. It is a space where engagement and collaboration are encouraged, and where spontaneous conversations are natural occurrences. While the bones of this decades-old building helped make such a conversion fairly streamlined, finding the right balance (and complying with the Living Building Challenge's high standards) wasn't so simple.

In its previous life as the Musser Library, the feet of the staircases on either side of the atrium faced the front entrance, leading patrons upward as they entered the building from Iowa Avenue. To subtly communicate the new nature of the adapted building, from public commons to a private place of work, the stairwells' direction was reversed a full 180 degrees, with the feet now facing the back of the building. (The old staircases were not retained, although many of their parts were salvaged for later use, which is addressed in both the Materials and Beauty Petals chapters.)

Within the larger atrium space, a composition of work and respite areas are designed at varying scales to accommodate virtually every need, mode, and size. Individuals can take a

seat under one of the staircases, just as a small team of people can take their lunch at the kitchen bar, have an impromptu meeting in the downstairs living room, or gather a larger group of colleagues at the central conference table. In each one of these scenarios and any other variation one can think of, people are physically connected to nature and natural elements.

Daylight now pours in from the three large skylights overhead. That light first touches the rooted Ficus Benjamina trees, which lend the Stanley Center a quirky yet palpable sense of standing inside a human-scale terrarium. This is further complemented by potted lilies, looming palm fronds, and entire walls of edible greens. Beyond the greenery, natural elements abound. From tectum wood fiber in the ceiling panels and oak treads for the stairwells, to stone and repurposed pearl button blanks providing unique finishes throughout, the interior design vision for the Stanley Center was to create an "organic living room." That vision, which has very much been achieved, is a gestalt of biophilic parts, goals, and principles. It is all the result of a design process that was not only grounded in this space, but responsive to the unique needs and wants of its employees and rooted in its community.

"The back-of-house rooms in the old library were classroom-type spaces, some offices, and staff resource areas. None of those spaces had windows, and no natural light entered the building from that side."

SALLY OBERNOLTE
Principal, Neumann Monson Architects

"With acoustics in mind between the new [private] offices, we had to figure out the wiring and keep them concealed within the stud cavities while dealing with logistical issues. This was quite challenging in terms of meeting the client's needs and managing acoustic aspects as well."

KHALID KHAN
Principal, Neumann Monson Architects

UNCOVERING LIGHT AND LANDSCAPE

Achieving the right balance of biophilic design in any professional setting involves more than designing with timber elements and introducing house plants into a space, even if some of them happen to be rooted to the ground. Biophilia goals vary from introducing natural elements and green space to providing vistas and blurring the boundaries between internal and external spaces.

On behalf of the Stanley Center, the team at Neumann Monson devised a complex Biophilia Matrix to analyze and determine the top design priorities for the organization's new space. Healthy air, light, wellness, and aesthetics were all key concerns. These come to bear through the soaring skylights, the use of plant pesticides chosen from the Organic Materials Review Institute's database, MERV 13 filters in the building's HVAC system, and by designing with wood and warm earth tones, among other features. In the end, however, the design team arrived at their top three goals:

1. Align the office space, operations, and daily workflows with the mission and vision of the Stanley Center.

2. Create acoustically sound spaces, both private and shared, that may contribute to a focused and collaborative work environment.

3. Provide a variety of spaces for work and collaboration, including comfortable and intimate spaces and large configurable spaces.

Neumann Monson's Brandon Fettes observes how the Stanley Center's biophilic design elements all work together as a curated and intentional sequence of elements. "As you walk past the landscaping along the building's perimeter and see the large cisterns on the corner, that's telling you a story of place and it's communicating part of the building's process. Then you come in through the vestibule and enter the living room, and you are met with different but connected landscapes. Beyond that is the kitchen and café area, which connects to the courtyard, bringing the interior outside and vice versa. The living room becomes the cafe which then becomes the zone where you draw out all that materiality, the warmth of the wood, the patterning of the stone. All these things give you the sense of being connected to nature."

Fettes and his colleagues also note how this combination of steps, turns, and abbreviated vistas, regardless of their exact sequence, comprise only the experience of entering the building. Light and landscape are part and parcel of the Stanley Center. They are integral components of each staff member's entry experience before they even reach their respective offices. But as the designers of this experience are keen to point out, chance encounters and collaboration are key to the Stanley Center's mission and vision, and now too of their workspace, where common spaces, long views of the atrium and the art-filled corridors, and connections to the outdoors are central components of their work environment.

"The courtyard was one of the biggest design solutions for achieving our daylighting goals and meeting the Health and Happiness Petal requirement."

SALLY OBERNOLTE
Principal, Neumann Monson Architects

"Although it was a library building, there wasn't inherent acoustic properties built into the building."

LYNDLEY KENT
Principal, Neumann Monson Architects

"The stairs used to come up towards the west. We flipped that around not only to designate a more private second level, so the public isn't immediately drawn upward, but to bring people [further into] the building. This solution obliges employees to cross paths, pass one another's offices, and have chance interactions while getting to where they need to go."

SALLY OBERNOLTE
Principal, Neumann Monson Architects

FIRST FLOOR PLAN

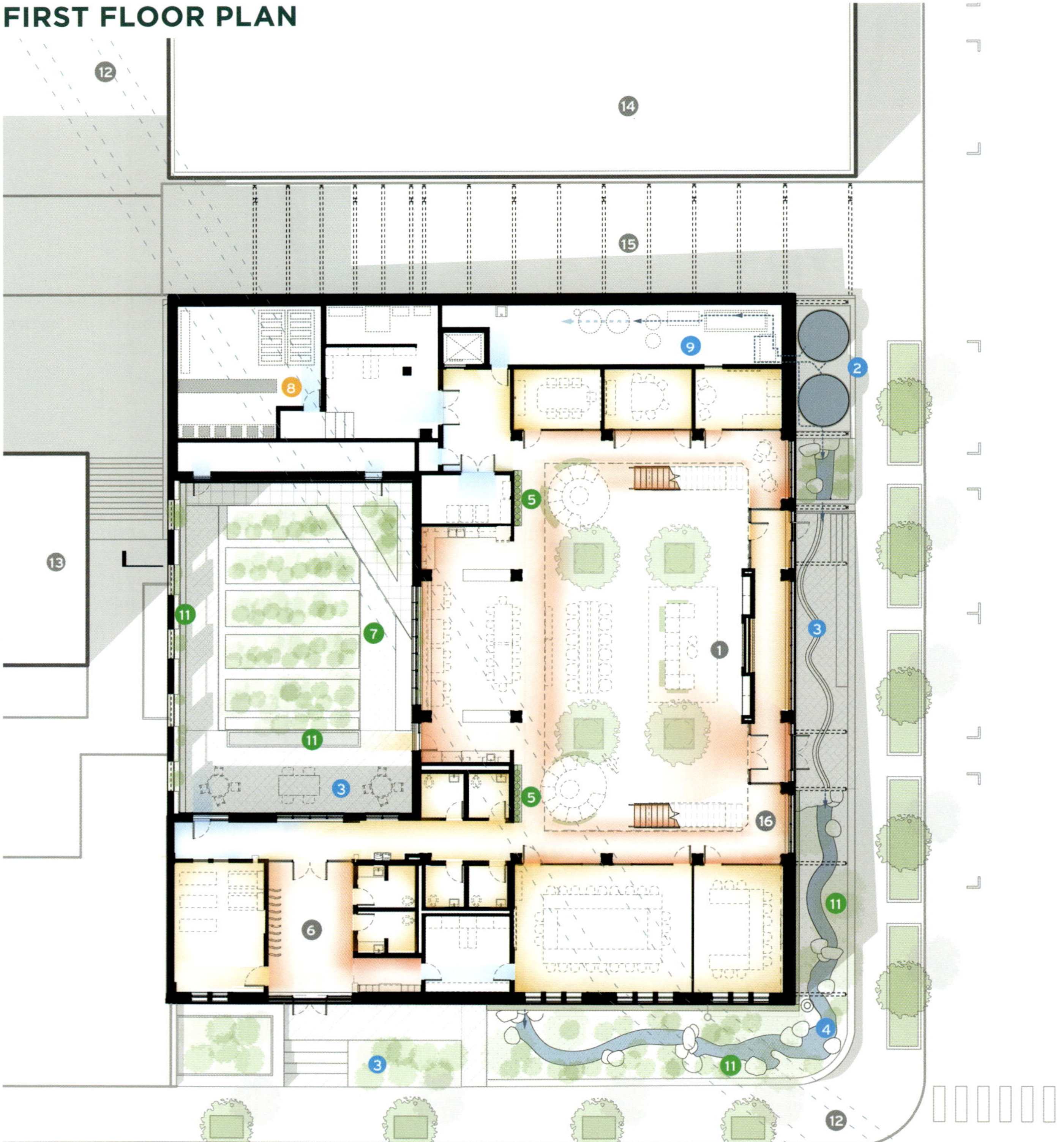

SECOND FLOOR PLAN

1. Living Room
2. Water Collection Cisterns
3. Permeable Paving
4. Storm Water Bioswale
5. Lettuce Walls
6. Mud Room
7. Urban Agriculture Plots
8. Solar Battery Storage
9. Water Supply Room
10. Office Green Wall
11. Native Plantings
12. Historic Sewer Line
13. Low Income Public Housing
14. Muscatine Center for Social Action
15. Alley (Food Pantry Access)
16. Tree of Accountability

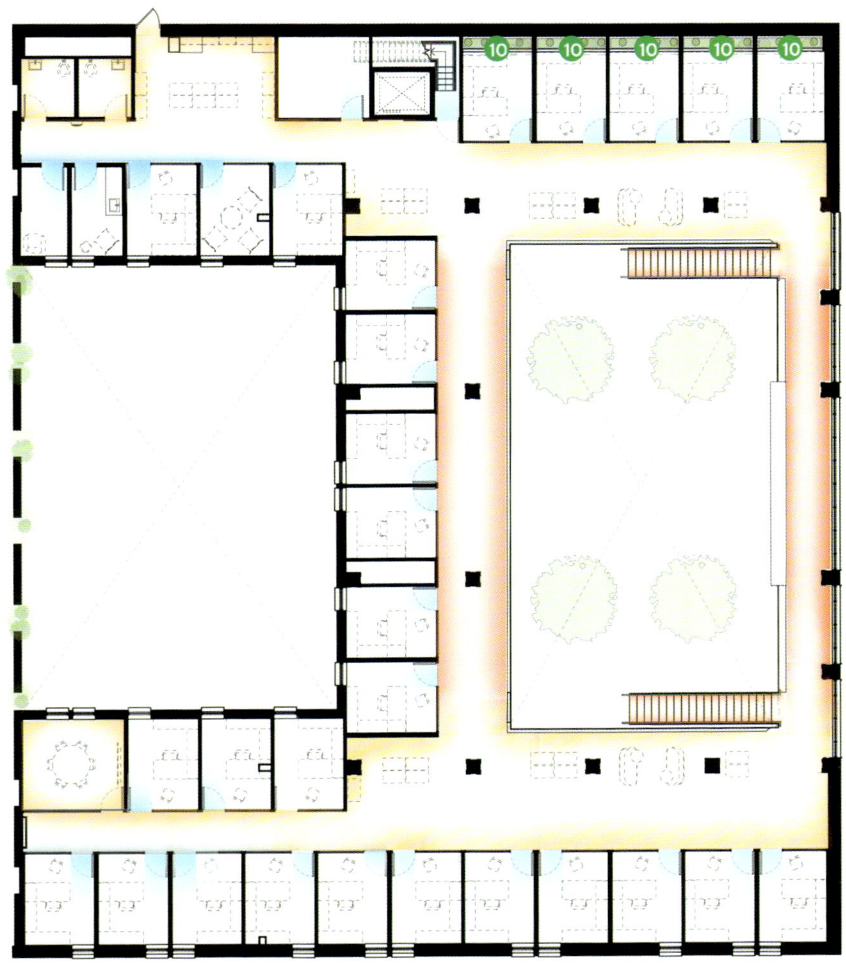

CIRCULATION

- Heavy Intensity
- Moderate Intensity
- Light Intensity

ACTIVATING EVERY CORNER

Within the nearly 24,000-square foot Stanley Center and the broader confines of the building's footprint, from its courtyard to the rooftop to the landscaped swales near the corner of Third and Iowa, nary a square foot of space is wasted.

When adapting and reusing a decades-old civic building with good bones but little architectural character, the opposite result could have easily been the case.

"Every corner of the building is susceptible to some portion of glare, but the site does a good job of helping with this issue," says Brandon Fettes. What also helps are the narrow, vertically oriented punched windows that run in a staccato pattern along the upper floor facing Third Street. In fact, except for the Stanley Center's front façade, almost every inch of the building's second floor perimeter is occupied by private offices.

In the era of hoteling, hybrid work arrangements, and open office plans, this may seem like an odd choice. But the Stanley Center is not your everyday organization. Considering the high level and sensitive nature of the work carried out by the foundation daily, providing its staff members with some degree of privacy was essential to maintaining successful workflows.

The team at Neumann Monson wanted to encourage movement with their design choices while also honoring their client's needs. Working with the building's existing window openings would have "admittedly worked better" and saved on costs, according to Sally Obernolte. "But because the Stanley Center felt strongly about everyone needing their own office with their own windows and natural light," configuring the existing openings to meet that need would have resulted in making "Swiss cheese" out of the southern façade and, consequently, the interior office layout.

Beyond the south-facing side of the building, it was important to "activate every corner," says Obernolte. This meant designing for amenities and creating points of interest, collaboration, and respite. It also meant crafting a larger office layout that obliged Stanley Center staff to touch several of these points en route to their respective offices. This is especially relevant for those staff whose offices are in places that might otherwise not receive a lot of foot traffic.

The living room nodes, green amenities, and banquettes of varying size are not limited to the first floor. Along the building's northern side, where reconfiguring window openings wasn't an option (there were no existing windows to begin with), private offices now benefit from individual skylights and green walls directly underneath, providing thriving backdrops for those employees. And just outside those offices, the open corridor overlooking the atrium features smaller living rooms that co-mingle with the Ficus trees as they ascend towards the skylights overhead.

Throughout the building, each common space is treated like a potential amenity space. This includes small seating areas that are open and visible yet discreetly tucked away. It extends to the several enclosed but windowed conference rooms along the edges that don't prevent natural light from entering the building but also help mitigate glare within the building's interior. And it very much applies to the Stanley Center's prominent southeast corner, where the new sculpture "Tree of Accountability" stands at attention on the ground floor and invites passersby to examine its distinct materiality, peruse its myriad leaves and written messages, and consider the work within the broader contexts of people, place, time, and history. (The artwork is discussed in further detail in the Materials and Beauty Petal chapters in PART III.)

The Health + Happiness Petal's Healthy Interior Performance Imperative specifies that 95 percent of regularly occupied spaces have access to daylight and outside views, while allowing opportunities for occupants in the remaining 5 percent to freely move to those compliant spaces. Virtually every corner of the Stanley Center benefits from abundant natural light and each one boasts its own distinct character. A diverse collection of huddle rooms, wellness rooms, and living rooms all intermingle with miscellaneous common areas and corridors to produce an unquestionably healthy and dynamic workplace environment.

"Every corner of the building is susceptible to some portion of glare, but the site does a good job of helping with this issue. Incorporating many of the green elements, both inside and outside, does some of the work. Nature also helps."

BRANDON FETTES
Intern Architect,
Neumann Monson Architects

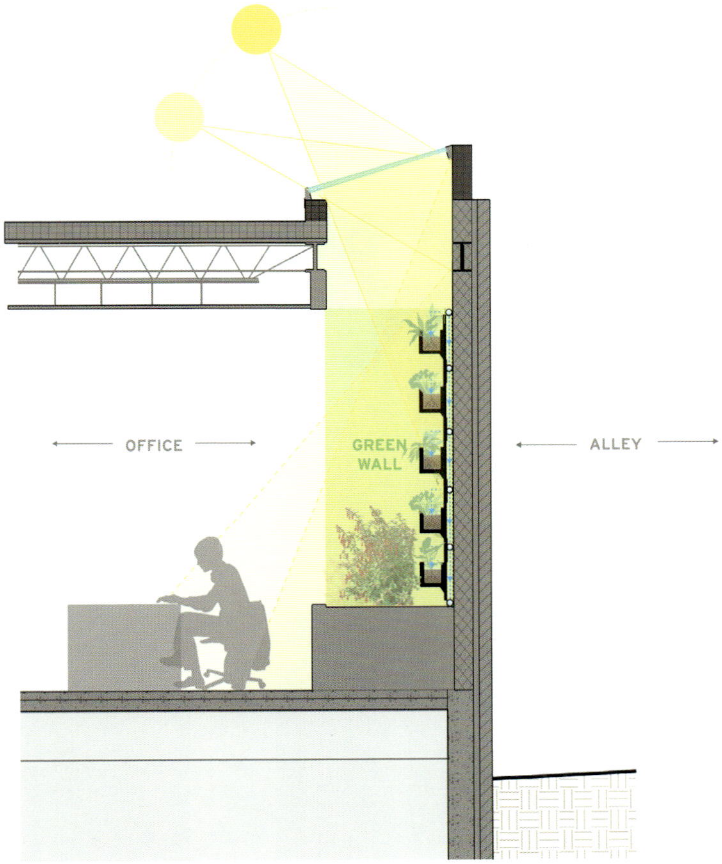

HEALTH + HAPPINESS OVERVIEW

- 95 percent of occupied spaces have daylight
- Four large interior trees and over 300 live plants
- Herb and vegetable garden on-site

THE MATERIALS PETAL

The Most Challenging Petal

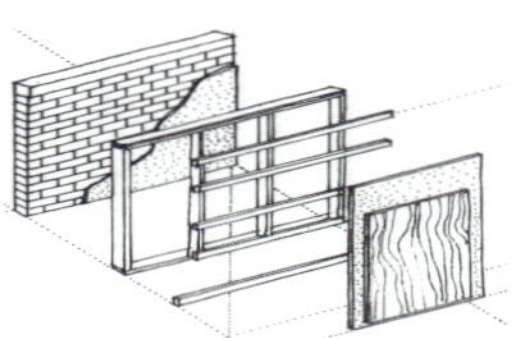

"Our job and the architects' job was basically to make one cohesive building out of two separate pieces. It was a big challenge, figuring out what was what, what was bearing down on what in terms of foundation, and making sure everything was in its right place."

BRAD HILL
Principal, Raker Rhodes Engineering

THE MOST CHALLENGING PETAL

When it comes to meeting the criteria and stringent benchmarks of the Materials Petal, this is where the rubber that is the Living Building Challenge meets the road.

"It has been *really* challenging," says the Stanley Center's CEO Keith Porter. "If people are going to falter in their pursuit of the Living Building Challenge, it will likely come down to building materials. I just hope that our process helps make it easier for others in the future."

Such hope is admirable. Success, however, is not as easy to come by. On average, the building materials supply chain comprises roughly 50 percent of a building's cost. And many of those raw and composite materials, like steel, concrete, glass, stone, brick, minerals, and timber, lack transparency in terms of their provenance, ethical labor practices, and base ingredients. Much of the building industry's global supply chain is shrouded in mystery, with few tools available to design and building professionals that enable them to confirm if certain materials were sustainably extracted and what, exactly, they are made from. And while not every component that goes into the building must come with a Declare label or be 100 percent free of Red List content, the Petal does place a high premium on having a diversity of products from various manufacturers, regional sourcing, and diverting most construction waste from landfills.

"There are not a lot of materials out there that meet the standards of the Living Building Challenge," says Brian Mrozek, the project manager with Graham Construction. "On most projects, drywall is drywall and metal studs are metal studs. It doesn't really matter where they come from. You just go in, slap on these materials, apply some high-quality finishes, and you're done. Well, here, we just can't do that." Mrozek cites the Covid-19 pandemic as being particularly disruptive. In its throes, when manufacturers of flooring, roofing, insulation materials and more were scrambling just to meet demand, an incoming request to perform due diligence that goes well beyond the status quo could seem intrusive. "But in a pandemic world we had to think about how the standards for procurement and schedules had changed, and then we had to think outside that box. We asked more probing questions of suppliers."

Graham Construction pouring mock-ups of permeable pavement, which was used under pavers throughout the project to allow water to soak into the ground

"Even before [our trade partners] arrive on-site, we have communicated to them that every single material has to be entered into our software to be vetted. There was a huge educational component ... It rocked the sub-contractors' worlds because they had to think so much further ahead than they normally would on a project."

GINGER ZENOR
Project Engineer, Graham Construction (formerly)

"There are not a lot of materials out there that meet the standards of the Living Building Challenge. On most projects, drywall is drywall and metal studs are metal studs. It doesn't really matter where they come from. You just go in, slap on these materials, apply some high-quality finishes, and you're done. Well, here, we just can't do that."

BRIAN MROZEK
Project Manager, Graham Construction

Mrozek's former colleague at Graham, Ginger Zenor, a Living Future Accredited professional who served as project engineer, cites the importance of communication and transparency when dealing with sub-consultants, particularly as it concerns vetting materials and ensuring strict protocols are in place. "Even before [our trade partners] arrive on-site, we have communicated to them that every single material has to be entered into our software to be vetted," Zenor says. There were some growing pains, she admits. Quick runs to the hardware store for a replacement part or simply tossing pieces of scrap metal into a dumpster are not options when constructing a Living Building. "There was a huge educational component" to making this building a reality, she says. "It rocked the sub-contractors' worlds because they had to think so much further ahead than they normally would on a project."

Such a stringent vetting and submittal process for every stud, every panel, every window, and every square foot of concrete poured and insulation installed, to cite just a few examples, can also dramatically impact scheduling, which was the case here. Many suppliers will instantly recognize the uniqueness of the circumstances they're entering and adjust accordingly. Despite the requisite hand wringing and, frankly, thinner profit margins for everyone involved, Zenor clearly understands the value of being associated with a project like the Stanley Center and communicated as much to Graham's partners. "There are benefits and there are growing pains," she says. "That's just the reality. So, we communicate to our partners the mission of the project and how it's different. And often, the suppliers get so excited that they'll move their production time up to meet our schedule. They figure it out."

Above all else, the "integrity of the project" is what brought everyone to the table and raised their game, according to Graham's leadership. "It was a huge group effort," Zenor says, "but it wouldn't have happened without everybody working together."

Once more, this notion of collaboration and chain reactions comes to bear. More so than even other demanding and materials-intensive Petals, like Energy and Water, the Materials Petal is often the most time-consuming. The reason for this is simple: designers, engineers, contractors, and other trade professionals are navigating seas that are relatively uncharted.

From the smallest metal fasteners to wholesale quantities of roofing insulation and everything in between, the mandate to thoroughly source, vet, analyze, record, procure, transport, and install all variety of building components that actively contribute to a non-toxic, ecologically restorative, carbon positive, and transparent materials economy, all while diverting as much excess material and waste as possible, is a truly monumental task. Only by assembling a diverse team of dedicated, highly communicative, and (in most cases) altruistic building professionals can such a task be fulfilled.

RED LIST COMPLIANCE

Unknown factors abound in the Living Building Challenge. Tackling the Materials Petal in a Living Building project means contending — in a best-case scenario — with what you know you don't know. The International Living Future Institute's Red List provides an invaluable guide to discovering just that, and then charting unique pathways to abiding by that high standard.

In keeping with the latest iteration (4.0) of the Living Building Challenge's Red List, which is now organized by chemical class rather than by specific materials or chemicals, project teams must vet a minimum of 90 percent of materials by cost against the Red List. These Red List changes are designed for teams to prioritize their vetting and research efforts and they recognize the limited documentation and availability of Red List-compliant products across all categories.

"The architects reaching out [to us] are asking that manufacturers and suppliers consider participating in programs like Declare or [the Health Product Declaration Collaborative]. We're a third-party consultant, so we just track down the information, analyze it, and make it available so that project teams can find comparable products. But in terms of meaningful advocacy, that needs to come from the product specifiers. Having those kinds of conversations across integrated teams is critical."

BRENNA WISSMAN
Project Manager, Integrated Eco Strategy

The project team, led in part by general contractor Graham Construction, began by removing asbestos from the existing building as well as hydraulic fluid from the old elevator shaft. With the building thoroughly gutted and voided of hazardous materials, Neumann Monson's team got to work on finding the right balance of new materials versus opportunities to upcycle existing ones that would contribute to the Petal's Net-Positive Waste Imperative.

For new products, the Petal's Responsible Sourcing Imperative requires one Declare product for every 100 square meters of gross building area, or up to 40 Declare products for buildings with a project area 4,000 square meters or more. Translating that for the Stanley Center's 1,588 square meters (19,823 square feet) of occupiable space, the minimum number of Declare labels would come to 16. The newly completed Stanley Center contains 63 Declare products!

Any way you measure it, meeting the 90 percent Red List requirement and the Declare requirement go hand in hand. "There is a direct correlation between [Declare] products and the Red List because [manufacturers] have to pledge to be Red List free," says Brian Mrozek. "So, get as many Declare products in there as possible and you get that much closer to your Red List goals. That was our approach."

Of course, tracking down Red List-free building materials and vetting their efficacy are tasks that don't exist in a vacuum. Like all other Petals, this design exercise, and the Imperatives it touches (e.g., Red List, Responsible Sourcing, and Living Economy Sourcing, as well as Imperatives from other Petals like Healthy Interior Performance and Beauty + Biophilia) is the product of close collaboration and transparency.

To ensure a thorough vetting process, the design team turned to Integrated Eco Strategy, a sustainability consultancy focused on healthy building materials. (The company has consulted on more than a dozen Living Building and Petal-certified projects.) The company maintains an industry leading crowdsource database called Red2Green, which features a library of roughly 20,000 products and growing, including comprehensive lists of product ingredients, manufacturer contacts, and VOC, FSC, and Red List status, among other details. With Integrated Eco Strategy's guidance, the integrated design team was able to specify dozens of Declare label products from the Red2Green database. However, as is the case with many Living Building Challenge projects, many whole building components that are critical for achieving high performance, like insulations designed for specific functions, do not yet benefit from transparency on the part of their manufacturers, or really any good faith attempts to void certain products of "worst in class" materials, chemicals, and other elements.

Integrated Eco Strategy's Brenna Wissman, a project manager with experience consulting on ten Living Building projects to date, cites three such products that required due diligence on their part to ensure suitable performance (while falling short of compliance) in lieu of actual documentation. The first of these is flexible elastomeric pipe insulation, a synthetic rubber with a closed-cell structure and smooth surface for vapor control; there are currently no versions of this product that participate in a standardized industry disclosure program. The second is FSK-faced duct wrap, a fiberglass insulation product. Wissman notes the availability of Red List-free unfaced options, yet the facing is required for vapor control, and the present-day market solution for meeting flame spread requirements in faced duct wrap is to add halogenated fire retardants (HFRs), a broad class of Red Listed retardants that contain bromine and chlorine. Finally, there is ASJ-faced pipe insulation, aka All Service Jacket, which is manufactured similarly to FSK facing and suffers from the same conditions.

In other instances, Wissman laments how some materials are Red List compliant but are in relatively short supply, or they happen to be part of a class of materials and finishes that aren't known for variety, at least in the compliance department. Until just a few years ago, there was no commercial grade interior paint that came with its own Declare label. Then in 2017, Benjamin Moore bucked that trend and achieved full compliance with its zero-VOC line of Ultra Spec 500 paints. (Benjamin Moore paint is on the walls of the Stanley Center.) And today, more and more paint manufacturers, from the high end to more accessible options, are beginning to follow suit. Wood stains and similar finishes, however, remain one area that lacks transparency and, by extension, compliance. Wissman not only sees this as a missed opportunity for the market, but also ironic given how prominent a role exposed timber and other wood products play in today's sustainably minded, high-performance buildings.

Declare.

ECOS Wood Stain
Imperial Paints, LLC

Final Assembly: Spartanburg, South Carolina, USA
Life Expectancy: 10 Year(s)
End of Life Options: Salvageable/Reusable in its Entirety

Ingredients:

Vehicle: Water; **Binder:** Methyl Methacrylate Polymer; Polymethyl Methacrylate; **Vehicle Purifier:** Sodium Hypochlorite; **Thickener:** Xanthan gum; **Corrosion Inhibitor:** Sodium Benzoate; Sodium Nitrite; **Surfactant:** Polyoxyethylene Stearyl Ether; **Preservative:** 1-Hydroxy-2(1H)-Pyridinethione, Sodium Salt; **Defoamer:** Tributylamine; **Antioxidant:** Polyoxyethylated Stearyl Alcohol

Living Building Challenge Criteria: Compliant

I-13 Red List:

■ LBC Red List Free	% Disclosed: 100% at 100ppm
☐ LBC Red List Approved	VOC Content: 0 g/L
☐ Declared	

I-10 Interior Performance: CDPH Standard Method v1.1-2010
I-14 Responsible Sourcing: Not Applicable

IPA-0007
EXP. 01 DEC 2024
Original Issue Date: 2015

MANUFACTURER RESPONSIBLE FOR LABEL ACCURACY
INTERNATIONAL LIVING FUTURE INSTITUTE™ living-future.org/declare

"There is a direct correlation between [Declare] products and the Red List because [manufacturers] have to pledge to be Red List free. So, get as many Declare products in there as possible and you get that much closer to your Red List goals. That was our approach."

BRIAN MROZEK
Project Manager, Graham Construction

The harsh reality is, there is no such thing as a building — at least on a commercial scale — that is 100 percent Red List free. Even abiding by the 90 percent standard demands greater accountability and due diligence across all phases of design, construction, operations, and maintenance than the building industry and its key players have ever been accustomed to. Every sustainable, high-performance, or so-called "green" building that is operating to this day contains some modicum of materials found on the Red List. This is no dirty little secret. On the contrary, it's a wakeup call.

The key to ascending that performance curve is advocacy, and given her experience in this arena, Wissman knows how important it is for manufacturers to play a larger role in public disclosure programs like the International Living Future Institute's Red List, Declare, or something similar. Only then will transparency become baseline, and consequently, a broader range and variety of common building materials will enter the market that are free of "worst in class" chemicals and other elements. "The architects reaching out [to us] are asking that manufacturers and suppliers consider participating in programs like Declare or [the Health Product Declaration Collaborative]," Wissman says. "We're a third-party consultant, so we just track down the information, analyze it, and make it available so that project teams can find comparable products. But in terms of meaningful advocacy, that needs to come from the product specifiers. Having those kinds of conversations across integrated teams is critical."

Declare.

Beyond Walls Stanley Center
Allsteel Inc.

Final Assembly: Muscatine, Iowa, USA
Life Expectancy: 30 Year(s)
End of Life Options: Salvageable/Reusable in its Entirety, Take Back Program (Anew), Recyclable (22-100%)

Ingredients:

Fumed Silica; Wood Fibers; Aluminum; Glass, oxide, chemicals; Urea, polymer with formaldehyde and 1,3,5-triazine-2,4,6-triamine[1]; Urea, polymer with formaldehyde[1]; Iron; Polymethylene polyphenyl isocyanate; Urea; Hydrocarbon waxes (petroleum), hydrotreated microcryst.; Phenol, polymer with formaldehyde[1]; 1,3,5-Triazine-2,4,6-triamine, polymer with formaldehyde[1]; Titanium dioxide; Quercus; Oak; Carbon black; Magnesium; Silicon; Zinc; Water; Copper; Undisclosed (0.031-0.0348%)[2]; 1-Propene, polymer with ethene; Manganese; Monoprene; Chromium, metallic; Titanium

[1]LBC Temp Exception RL-009 - Formaldehyde
[2]LBC Temp Exception RL-004b - Proprietary Ingredients in Declare

Living Building Challenge Criteria: Compliant

I-13 Red List:

☐ LBC Red List Free	% Disclosed: 99.97% at 100ppm
■ LBC Red List Approved	VOC Content: Not Applicable
☐ Declared	

I-10 Interior Performance: CDPH Standard Method v1.2-2017
I-14 Responsible Sourcing: Product Available with FSC Chain of Custody

HNI-0001
EXP. 01 JUL 2024
Original Issue Date: 2022

Third Party Verified

MANUFACTURER CLAIMS VERIFIED BY WAP SUSTAINABILITY
INTERNATIONAL LIVING FUTURE INSTITUTE™ living-future.org/declare

"As the contractor, we looked local first and tried to work with companies that were as close to Muscatine as possible. There were times when a trade partner was too small and didn't have the necessary staff, and some were just intimidated by the project. But in all we tried to keep local."

GINGER ZENOR
Project Engineer, Graham Construction (formerly)

"We had supplier surveys and a lot of back and forth. We have a corporate [transparency] goal, and this project helped restructure our efforts to achieve that goal in a different way. It accelerated those efforts and it accelerated our understanding of what it takes to build a product out."

LISA BRUNIE-MCDERMOTT
Director, Corporate Social Responsibility, HNI Corporation

ALL IN THE FAMILY

The Stanley Center for Peace and Security owes its existence, in part, to another Max Stanley venture, the office furniture manufacturer HNI Corporation, which he co-founded in 1944 with Clem Hanson, an advertising executive, and H. Wood Miller, an industrial designer.

Originally established as Home-O-Nize, the company's success and growth over the ensuing decade allowed Max and Betty Stanley to use shares of company stock (160 shares at $10 per share) to formalize the Stanley Center, in 1956.

Today, the HNI Corporation — which includes multiple well-known subsidiary brands, such as The HON Company, Allsteel, Gunlocke, HBF, and Design Public Group, among others — remains headquartered in Muscatine, on the banks of the Mississippi River, and has grown into one of the largest manufacturers of workplace furniture in the world. And it was HNI that supplied the Stanley Center with furniture and furnishings for its new headquarters, including numerous Allsteel and Gunlocke products.

"As the contractor, we looked local first and tried to work with companies that were as close to Muscatine as possible," says Ginger Zenor. "There were times when a trade partner was too small and didn't have the necessary staff, and some were just intimidated by the project. But in all we tried to keep local." Even if sourcing every material in service of a local living economy wasn't possible, it was fortuitous that for one of the building's most demanding components — the furniture — the Stanley Center could look to a company with the means, wherewithal, and motivation to be part of a Living Building. What's more, that company shared a common lineage with the Stanley Center, and it just happened to be located down the road.

Prior to the Stanley Center making the choice to pursue a Living Building, HNI did not have any established protocols or practices that specified Red List materials in its product lines. "We never had an undertaking like this previously," says Bryanna Johnson, a product application manager with HNI. That's not to suggest, however, that the corporation didn't prioritize transparency in some manner before the Stanley Center came calling.

In fact, HNI had an existing goal — established in 2018 — to achieve 100 percent material transparency by 2025. And it was HNI's involvement with the Stanley Center project that accelerated that targeted benchmark, according to Lisa Brunie-McDermott, HNI's director of corporate responsibility. "We restructured our practice to be able to achieve what we needed to do for this project," she says, and it "gave us a vision of the scale required" to transition two noteworthy product lines into the black.

The products in question are Allsteel's signature Beyond architectural wall systems and Gunlocke's Silea case goods (desks, storage, and related items), which both earned Declare labels in 2022, and both of which have found a home in the Stanley Center. "Previously, we were just trying to get chemical transparency information for different categories of materials, which isn't easy," Brunie-McDermott says. "Certain materials are easy," she admits, "like wood and metals because you can use the American Society for Testing and Materials (ASTM) standard

"The Living Building Challenge is less about building these projects and more about the effects they have on the industry."

LYNDLEY KENT
Principal, Neumann Monson Architects

and access information that is publicly available. But to do a full product, we had to rearrange our scope of focus."

That kind of comprehensive analysis and vetting process required HNI's team to be in constant contact with their supply chain partners, "educating them on why we're asking for this information and why it's important to keep to the timeline that we were under," Brunie-McDermott recalls. "We had supplier surveys and a lot of back and forth. We have a corporate [transparency] goal, and this project helped restructure our efforts to achieve that goal in a different way. It accelerated those efforts, and it accelerated our understanding of what it takes to build a product out."

Indeed, vetting the chemical framework of a single material is not too tall a task, relatively speaking, and is something HNI had prior experience with thanks to its in-house database. A full product and its myriad components, however, can comprise several thousand lines of chemical information, as opposed to a couple of dozen. Further, adjusting the corporation's "design for environment" process to include not just product design but into something that caters to customer choice and layout, and with all the transparency expectations that come with it, demanded greater scrutiny and data analysis than anything HNI had engaged in prior. For that reason, Brunie-McDermott and her colleagues reflect on a process that was "educational," "evolving," and "not stress free."

HNI's involvement with the Stanley Center is not only indicative of a team (and family) effort to bolster living and regional economies, but to further the types of chain reactions that are critical to the Living Building Challenge's continued growth. A change in one material choice or building component can have ripple effects throughout a project, impacting not just one but several Petals and Imperatives. The same condition applies to one company's decision to realign its practice by instituting greater accountability and transparency.

Top: Table made by local craft person from felled ash from area.
Bottom: Button blank backsplash shadow boxes use no grout so buttons can be removed for reuse/historic preservation.

In HNI's case, the corporation is currently revamping its database of banned materials to be more comprehensive, in keeping with the Red List. It has earned Declare labels for two of its most popular and lucrative product lines. And it has committed to integrating transparency into its broader supply chain. This means that, going forward, the suppliers and consultants partnering with HNI will likewise be held to higher standards. This revamped supply chain strategy will now empower the corporation to score and assess suppliers on ESG-related topics and to establish a baseline tally that can be applied to a "best in class" listing of qualified and conscientious companies that operate within the supply chain.

"Our choices are governed by more than just aesthetics," Brunie-McDermott says. "I think we're getting there. It's a journey, and we're not yet where we want to be. But we've made progress."

RESPONSIBLE SOURCING AND SALVAGING

Some architects are aesthetes, pure and simple. This has been so ever since human civilization first laid brick and stone, and it will surely continue for millennia to come. Only now, at this moment, we find ourselves at an inflection point, where our preoccupation with a building's performance has gained equal footing with its purpose.

Beautiful buildings have always been designed with grand purpose, whether to celebrate community, the arts, a deity, or some other civic function that one culture or another deems worthy of monument. And while the Living Building Challenge does demand beauty with purpose, its mandate for high performance constitutes a higher calling. And that calling isn't to celebrate itself, but to motivate others to do better. According to Lyndley Kent, an architect with Neumann Monson, "The Living Building Challenge is less about building these projects and more about the effects they have on the industry."

The desired effect is to normalize accountability and transparency, and to establish today's best-in-class standards as tomorrow's baseline. That will require scaling up to absorb the added costs associated with the stringency of the Living Building Challenge. And to scale up, more industry stakeholders must be actively engaging in practices of responsible sourcing, materials conservation, net-positive waste management, and salvage.

Interestingly, strictly from a materials standpoint, the cost differential between standard and sustainable materials isn't especially noteworthy, according to Neumann Monson's Sally Obernolte. "It doesn't necessarily cost more in materials to do [a Living Building]. But it costs a lot more in time and effort!"

A big piece of that additional time and effort stems from the need to diversify a project's sub-contractors, manufacturers, and vendors, and ensure that a Living Building project is incorporating "place-based solutions" and contributing "to the expansion of a regional economy rooted in sustainable practices, products, and services." What this means, in so many words, is that Living Buildings are meant to spread the love (and knowledge) and not be created in a vacuum by just a few experts with proprietary IP.

One noteworthy change between previous iterations of the Living Building Challenge (3.1) and the current version (4.0) is that project consultants coming from within 2,500 kilometers of a project location is no longer a requirement. This change occurred to "encourage the sharing of expertise across project teams." That said, the Stanley Center is largely the product of a living and regional economy, with a great number of local and regional vendors contributing to the final product.

FSC-certified lumber and plywood for portions of the Stanley Center's interior fit out — used primarily for blocking and sheathing — were supplied by Certified Wood Products, based in Maple Lake, Minnesota. (Of note, the same company was a subcontractor for the Wolf Ridge Environmental Center, in Finland, Minnesota, the very first renovation project to earn full Living Building certification; the Stanley Center is striving to be the second to earn that distinction.) "The assumption on some of these green building programs is that the closer the resource extraction and processing is to the job site, the more environmentally friendly it is. And that's just not the case," says Dan Haugen, president of Certified Wood Products. "Just because something is close doesn't mean it was responsibly or sustainably extracted and processed."

Opportunities to employ materials and supply chains that are within close proximity to a project are increasing with each passing year, regardless of location, for the most part. As Graham Construction's Brian Mrozek recalls, "You can't just run down to Menards or the local lumberyard to get fire-treated board, because we need the documentation to prove what we're doing" is by the book, that it's a "high quality wood" with the necessary certification.

Other examples of high-quality wood can be found throughout the Stanley Center, both new and salvaged. The engineered Accoya panels used for the building's exterior siding, charred using the Japanese technique of Shou Sugi Ban (aka Yakisugi, or "burned cedar"), were sourced directly from a manufacturer in Japan. To ensure there was zero waste on the job site, in keeping with the Materials Petal's Net-Positive Waste Imperative, off-cuts from the Accoya panels were upcycled and used for much of the Stanley Center's interior signage.

MATERIAL AXONS

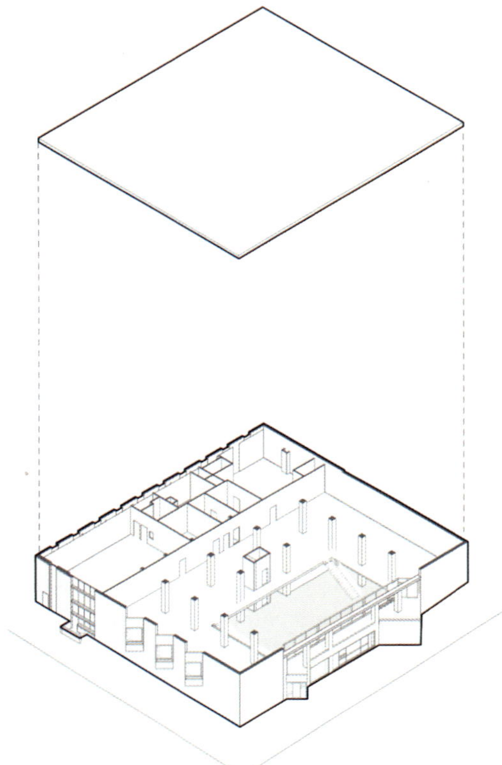

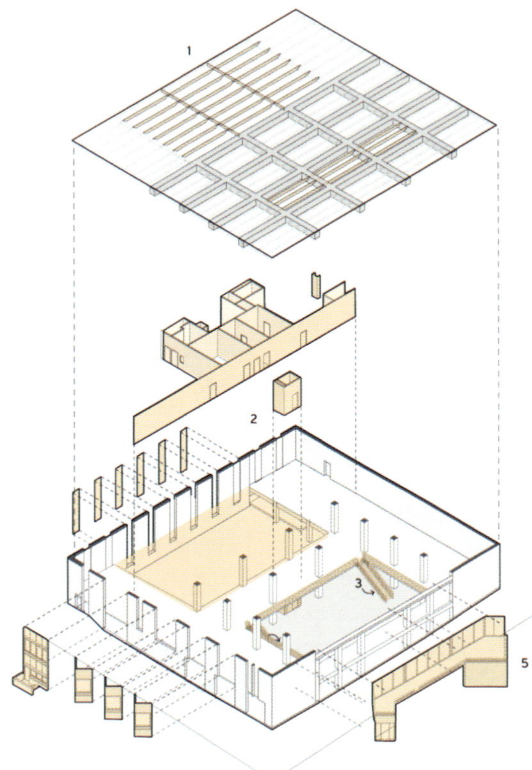

EXISTING

SUBTRACTION

1. Created openings in structure
2. Removed interior walls and floor slab
3. Removed and repositioned interior stairs
4. Utilized existing coffered ceiling for skylights
5. Removed exterior extrusions
6. Outdoor garden space

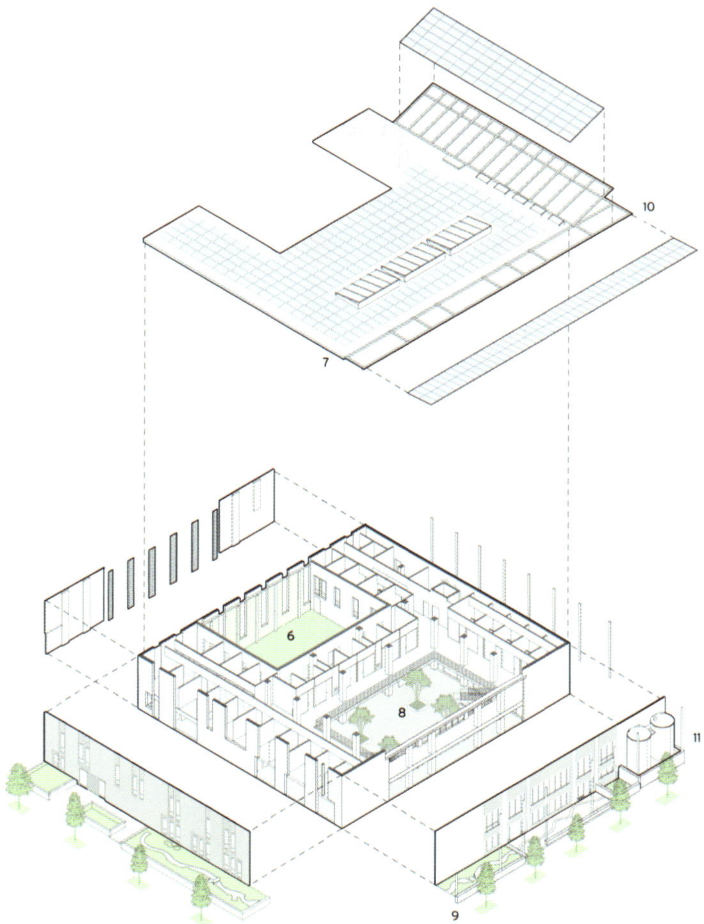

ADDITION

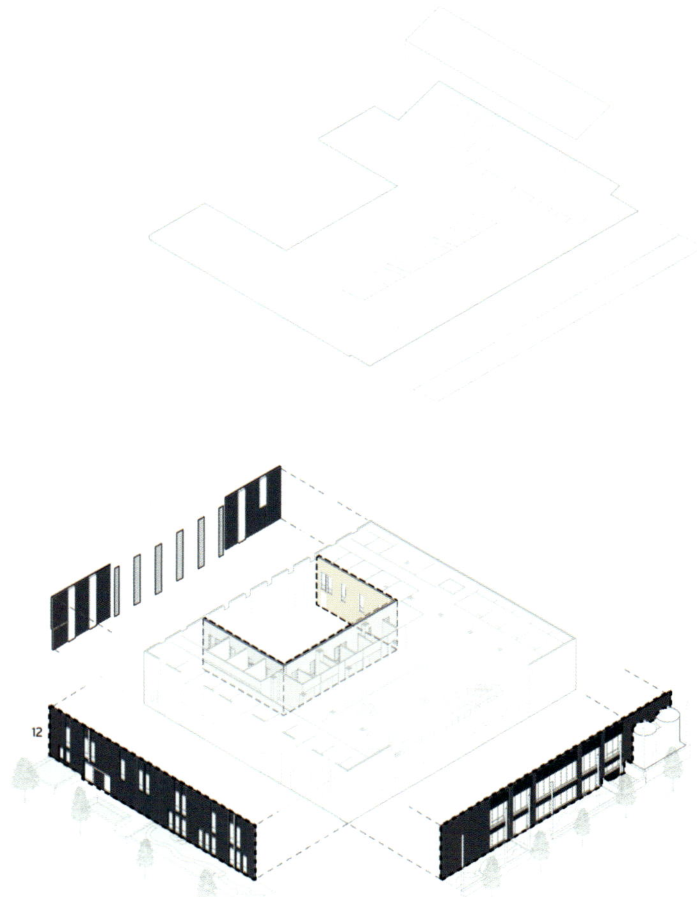

ACCOYA RAINSCREEN SYSTEM

7. PV-Panel on roof
8. Biophilic living space
9. Landscaping buffer at street
10. Canopy to extend PV
11. Rainwater cisterns
12. Accoya rainscreen system

BUILDING REUSE

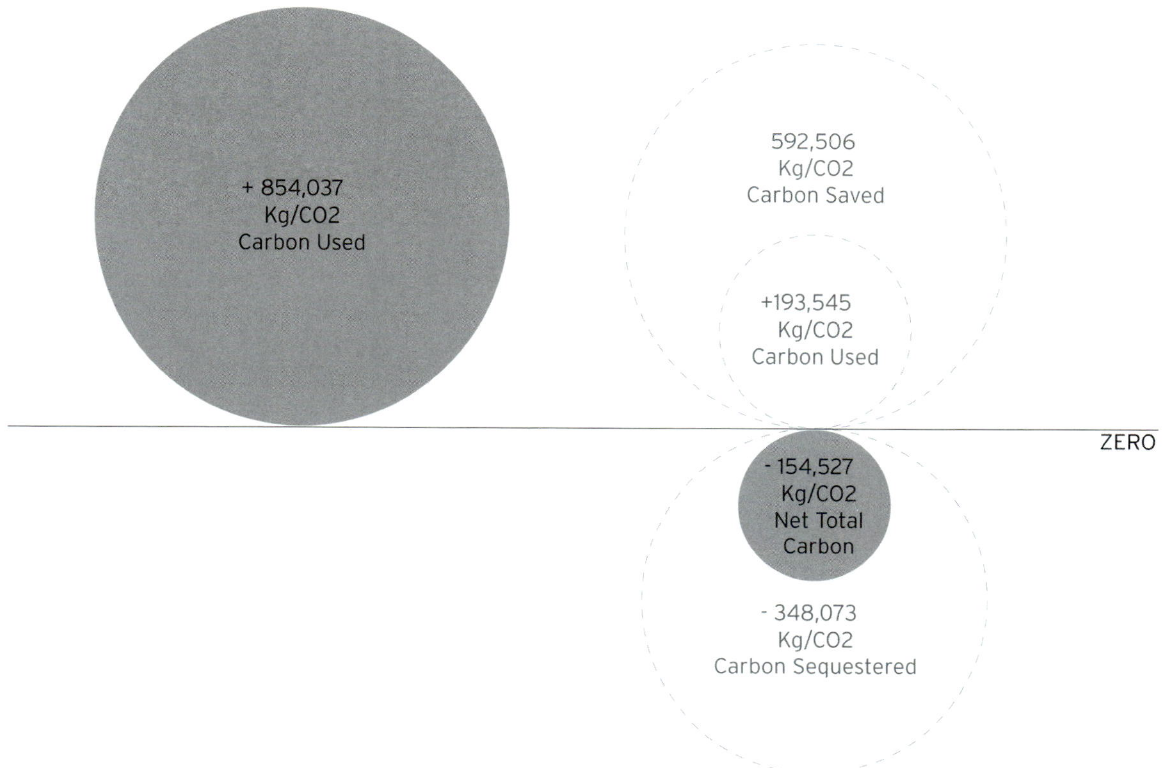

BASELINE NEW BUILDING **STANLEY CENTER PROJECT**

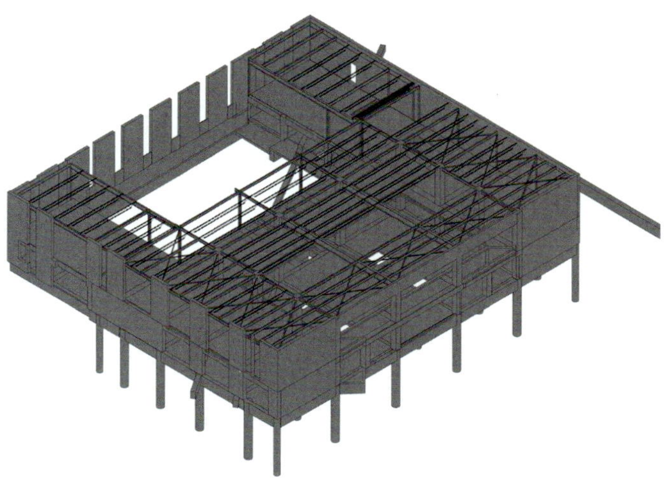

SALVAGED STRUCTURE

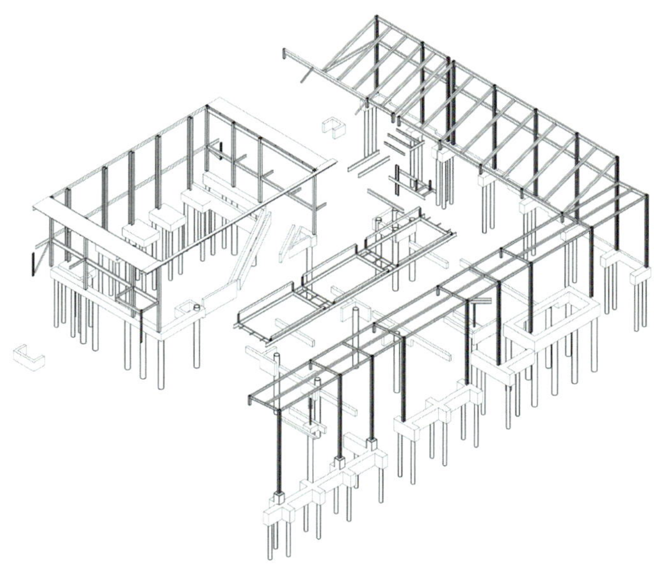

NEW STRUCTURE

500km
48%/25%
1,000km
16%/25%
5,000km
27%/30%
Muscatine, IA
BOLD = Project Results
LIGHT = LBC Requirements

> *"All the things that we don't know about what's in the built environment around us is extremely humbling, and a little scary. It has been really interesting to see how that plays out in a practical sense, with the types of decisions we have to make — that we get to make — about what we're putting into our building."*
>
> **CATHERINE ELLIOTT**
> Member of Governance,
> Stanley Center for Peace and Security

> *"The assumption on some of these green building programs is that the closer the resource extraction and processing is to the job site, the more environmentally friendly it is. And that's just not the case. Just because something is close doesn't mean it was responsibly or sustainably extracted and processed."*
>
> **DAN HAUGEN**
> Certified Wood Products

Often with Living Building projects, a lot of time and effort is devoted to just procuring a single material or piece of furniture, and the Stanley Center is no exception. For the building's new conference table, a pivotal and stunning piece that centrally resides in the ground floor atrium, the integrated team of Neumann Monson and HNI worked with Muscatine-based craftsman and furniture maker Matthew Johnson, owner of M.L. Johnson Woodworks, to manufacture a sustainable and functional statement piece. Johnson sourced six slabs of local Ash lumber — newly felled due to destruction caused by the invasive and non-native emerald ash borer — from Johnson Creek Hardwoods (no relation), a mill in Mount Carroll, Illinois, roughly 90 miles upriver from Muscatine. The slabs all came from the same tree, and were planed, jointed, sanded, cut, and fit by hand, "while leaving as much of the natural surface defects intact as possible," Johnson says, including several divots and crevices from where ash borer larvae embedded themselves — a stark but oddly beautiful reminder that nature's forces aren't always kind. Finally, Johnson finished the combined slabs with Oil Plus 2C, produced by Rubio Monocoat, a plant-based wood finish that is VOC free. "I've been using the product over the years, and it has become my preferred finish," Johnson says. "It is very safe to apply, since my shop shares the same air as my house, and it's safe for the people using the tables."

Many examples of salvaged materials and upcycling are hiding in plain sight, notably in the form of the pearl button blanks in the kitchen and the aluminum and decommissioned library book remnants that were used to create "Tree of Accountability." In other instances, salvaging efforts were a matter of waste diversion rather than creative reuse. Historic brick pavers from the former Musser Library site that were removed for site excavation and irrigation — principally in service of the Water Petal — were saved and are now in the possession of the city for eventual reuse.

At this evolutionary juncture in the Living Building Challenge, knowing what you don't know is the desired starting point for most project stakeholders. It is often the best we can hope for. But even getting to that point can be an arduous journey of self-discovery and harsh reality checks, particularly as it concerns building materials. According to Catherine Elliott, "All the things that we don't know about what's in the built environment around us is extremely humbling, and a little scary. It has been really interesting to see how that plays out in a practical sense, with the types of decisions we have to make — that we get to make — about what we're putting into our building."

MATERIALS OVERVIEW

- 48 percent of building materials came from within 500km of the job site
- 485 tons of reused materials
- 948 tons of waste diverted

THE EQUITY PETAL

Becoming A Just Organization

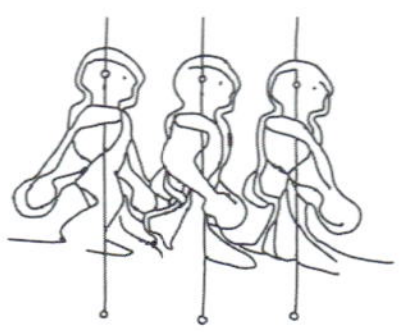

> *"We understood where the alignment was with our values, but it also gave us something to work towards."*
>
> **JENNIFER SMYSER**
> Vice President and Director of Policy Programming Strategy, Stanley Center for Peace and Security

BECOMING A JUST ORGANIZATION

Inclusivity in the workplace is about creating an environment that is safe and empathetic. It is about tethering an organization to its community, and fostering opportunity and growth through well considered hiring, employee policies, purchasing, and educational practices.

But like greenwashing a building's sustainable performance metrics with vague marketing speak, it's easy to talk big without much follow through. The International Living Future Institute's Just label is an innovative self-assessment tool that strives to reverse that behavior. Described as a "nutrition label for socially just and equitable organizations," the Just program invites organizations to be transparent and honest about where they excel and where they fall short across several social justice indicators: Diversity and inclusion, Equity, Employee benefits, Stewardship, Employee health, and Purchasing and supply chain.

"The Just label was very important to us. It's one of those things you don't really think about when starting a building project, because you might have to revise your employment policy," says Stanley Center CEO Keith Porter. "But it lined up so well with the Stanley Center's values, and it helped connect our building to the larger narrative that our governance has been promoting for a long time."

In no uncertain terms, the Equity Petal requires that at least two project teams with an integral role in decision making throughout design and construction have a current Just label, with an additional five teams, if applicable, needing to complete a Just Self-Assessment, or what amounts to a self-applied pre-screening. In addition to the Stanley Center for Peace and Security, Neumann Monson Architects, Integrated Eco Strategy, Environmental Consulting & Technology (ECT), Graham Construction and Biohabitats are all current Just organizations. HNI Corporation and Design Engineers both completed the Just Self-Assessment.

When considering his organization's journey of late, Porter grapples with what defines an equitable place of work. "The Just handbook gave us a set of definitions; it helped us benchmark ourselves. Of all the Petals, I would say Equity had the most positive social impact on our organization."

The Stanley Center's Jennifer Smyser, vice president and director of policy programming strategy, calls their approach to pursuing the Just label "considered" and "deep." "We understood where the alignment was with our values, but it also gave us something

"Going through this requirement has been remarkable. The Just concept is so brilliant, because we're benefitting as a firm, the project is benefitting, and our communities are benefitting."

DENISE SCHMITZ
Marketing Manager, Neumann Monson Architects

"The Just label was very important to us. It's one of those things you don't really think about when starting a building project, because you might have to revise your employment policy. But it lined up so well with the Stanley Center's values, and it helped connect our building to the larger narrative that our governance has been promoting for a long time."

KEITH PORTER
CEO and President, Stanley Center for Peace and Security

to work towards," she says. In that regard, the Just label is both a compelling tool for self-reflection and a roadmap for improvement. In one such example, Smyser cites her organization's new living wage policy that "ensures that no one is falling below the level of what it actually takes to live in our community. Not just a minimum or market wage, but a wage that allows them to live in Muscatine. This is more than the building influencing what we're doing; it's the building giving us those opportunities to walk the talk in so many different ways."

For others on the project, the Just label was an opportunity to look inward and reconcile how their stated values aligned with their practices; if the walk matches the talk, so to speak. It can be a sobering experience for some, particularly as it concerns gender and ethnic diversity, gender pay equity, benefits like family and medical leave, and employee wellness. The self-assessment process is designed to accurately reflect an organization's true self, blemishes and all. Neumann Monson wanted to get started right away.

Time is not typically a luxury on a Living Building project. Schedules are tight, supply chains are tighter, and the processes for everything from vetting materials to waste diversion demand constant attention and documentation. When it comes to pursuing the Just label, however, companies can bide their time, at least a little bit. They can peruse the handbook, examine the interrelated benchmarks, and begin to improve certain internal practices before performing any formal self-assessment. This can certainly help. But like starting to floss a week before going to the dentist, it's not necessarily doing any harm, but it's not making up for all past shortcomings either. The team at Neumann Monson didn't want to fudge their own numbers; they wanted to assess themselves as is, because they recognized that the label — much like the Living Building Challenge as a whole — has a ripple effect that extends beyond any one company's culture or any one building's footprint. The label should reflect everyone's and everything's true nature, so that earnest attempts to improve one's culture and practice is built on a foundation that is genuine, not artificially manipulated.

"Being propelled through this requirement has been remarkable," says Denise Schmitz, marketing manager with Neumann Monson. "That's what makes the [Just] concept so brilliant, because we're benefitting as a firm, the project is benefitting, and our communities are benefitting." Starting in the spring of 2023, Neumann Monson began updating its Sustainability Action Plan, in accordance with the AIA 2030 Commitment and the 2030 Challenge. This process led the team to question whether they should start requiring all their consultants to, at minimum, take the Just Self-Assessment. While the firm has not yet elected to codify this practice, it has become ingrained in their culture. "If we require it, will it motivate those firms to get their Just label?" asks Lyndley Kent. "This is something we've already been encouraging to many of our consultants whom we work with frequently."

The chain reactions stemming from the Just label are palpable for everyone involved, even for those who have engaged in the process but have yet to secure the moniker of Just organization. According to Ginger Zenor, formerly a project engineer with Graham Construction and a current LFA (Living Future Accredited) professional, Graham made it a requirement for all their trade partners on the Stanley Center project to take the Self-Assessment. "They didn't have to share the results with us, they just had to acknowledge that they'd taken it. Everybody can benefit from doing that internal check. It can serve as an eye opener."

JUST ORGANIZATIONS

Stanley Center FOR PEACE AND SECURITY

Biohabitats

ECT

GRAHAM CONSTRUCTION COMPANY

integrated ecostrategy

NEUMANN MONSON ARCHITECTS

JUST SELF-ASSESSMENTS

(2) Design Partners

(21) Construction Partners

Just.

Organization Name: Neumann Monson Architects
Organization Type: Architecture
Headquarters: Iowa City, Iowa
Number of Employees: 43

Social Justice Indicators:

Diversity & Inclusion
■□□□ Gender Diversity
■■■■ Ethnic Diversity
■■■■ Inclusion
■■■■ Engagement

Employee Benefits
■■■□ Health Care
■■■□ Retirement Provision
■■□□ Family/Medical Leave
■■■■ Training/Education

Equity
■■■■ Full-Time Employment
■■■■ Pay-Scale Equity
■■■□ Freedom of Association
■■□□ Living Wage
■■■■ Gender Pay Equity

Stewardship
■■■□ Local Communities
■■■■ Volunteering
■■■□ Animal Welfare
■■■■ Charitable Giving
■■■□ Positive Products

Employee Health
■■□□ Physical Health
■■■■ Well-Being

Purchasing & Supply Chain
■■■□ Equitable Purchasing
■■■□ Supply Chain

THE SOCIAL JUSTICE LABEL 2.0
NMA-002 EXP. 12/01/2024

INTERNATIONAL LIVING FUTURE INSTITUTE℠

Just.

Organization Name: Stanley Center for Peace and Security
Organization Type: Nonprofit
Headquarters: Muscatine, Iowa
Number of Employees: 26

Social Justice Indicators:

Diversity & Inclusion
■■■■ Gender Diversity
■□□□ Ethnic Diversity
■□□□ Inclusion
■■■■ Engagement

Employee Benefits
■■■□ Health Care
■■■□ Retirement Provision
■■□□ Family/Medical Leave
■■■■ Training/Education

Equity
■■■■ Full-Time Employment
■■■■ Pay-Scale Equity
■■□□ Freedom of Association
■■■□ Living Wage
■■■■ Gender Pay Equity

Stewardship
■■□□ Local Communities
■■■■ Volunteering
■□□□ Animal Welfare
■■■■ Charitable Giving
■■■■ Positive Products

Employee Health
■□□□ Physical Health
■■■■ Well-Being

Purchasing & Supply Chain
■□□□ Equitable Purchasing
■□□□ Supply Chain

THE SOCIAL JUSTICE LABEL 2.0
SCP-001 EXP. 10/01/2024

INTERNATIONAL LIVING FUTURE INSTITUTE℠

ACHIEVING ACCESSIBILITY

A sloping site does not necessarily lend itself to managing accessibility issues with ease. Complicate matters further with a building that is wedged into the corner of two converging slopes, as is the case with Third Street and Iowa Avenue, and the site planning challenge is compounded.

Neumann Monson's Sally Obernolte calls it a "tricky balance" of enabling access along the building's southern and eastern faces, and "lining up those points with public ways to ensure there are nice, flat entrances on both sides."

Coupled with irrigation planning, which included the introduction of tree beds, mini bioswales, and permeable concrete, as well as the significant task of running new pipes underneath the existing storm sewer to conjoin the site's low-lying bioretention areas, the building's primary means of egress along the public-facing sides had to be leveled off within this complex collection of linear equations. Doing so wasn't easy, but manageable. Of course, the sea of intersecting slopes does not end with the front and side doors. Nearing the courtyard and the site's northwest corner, the presence of the underground storm sewer presented a whole new set of design challenges in terms of urban agriculture as well as personal mobility and accessibility.

"Everyone has the exact same style and size of space. They all have natural daylight. The building is organized to create more private, contemplative spaces on the second level, where private offices are located, and more public collaborative spaces on the first level. This stitching together of different program elements helps activate the entire building."

SALLY OBERNOLTE
Principal, Neumann Monson Architects

The site's grade illustrates the challenge of achieving accessibility for the adapted building

"We wrestled with how to approach what this would look like," says Lyndley Kent. The concrete slab that comprises the courtyard's elevated terrace overlooking the main garden bed also happens to be the same elevation for the floor of the Stanley Center's battery room. Tucked into a discreet corner along the building's northern edge, the solar battery storage room is a few steps up from ground level and housed in an oblong space with low ceilings (under six feet) within this slab. Efficiently located though it is, it's also a room that most staff don't need to access. Consequently, it is the only room or space in the entire building that is not universally accessible.

A multi-purpose wellness room on the building's second floor overlooks the courtyard

Another key consideration, given the site's collection of odd angles and slopes, was the building's elevator. The former Musser Library building, principally constructed in 1971, was by no means designed to comply with American Disabilities Act (ADA) or Architectural Barriers Act (ABA) standards. Nor could the library's old elevator be repurposed in any way, given its size and age, which likewise didn't meet ADA requirements. So, an entirely new vertical corridor had to be constructed, along with a new elevator with updated specifications to meet code and accessibility requirements.

The myriad considerations surrounding accessibility presented an equal number of design challenges. That said, the Stanley Center's new headquarters is a private building. It is a place of work, but also a place for learning, which is very much in keeping with the building's original use as a library and civic asset. "They want to invite the public in," Obernolte says, but within reason. Renting out conference rooms, lounging in the common areas, or taking one's lunch in the courtyard are not services available to the public. The building itself, however, represents one large opportunity — or rather, a composition of opportunities — to educate on the purpose, function, mechanics, synergies, and practical challenges of designing, operating, and maintaining a fully regenerative Living Building.

SUSTAINABLE ACTION AND SOCIAL JUSTICE

When the Stanley Center formed its Office Space Steering Committee, among the focus teams that reported to this committee was one devoted to public education. Despite being a private building, it was important for the Stanley Center as a mission-driven organization — and one poised to make its home in such a high-profile manner — to actively engage with the community and develop an understanding of how the people of Muscatine, and the Indigenous peoples whose ancestors called this land home, might benefit from this building.

The prevailing questions were: what kind of public component is appropriate, nay, necessary, for a building such as this? And what are the needs of these communities and how might the Stanley Center accommodate them?

Early in the process, prior to the Stanley Center even purchasing the property at 304 Iowa Avenue, the Public Education focus team initiated a lengthy community survey. Approximately 120 respondents engaged with such queries as:

- What are the best things about living and working in Muscatine?
- What aspects of the Muscatine area's history and culture are important to you?
- Is there anything about the area you wish you knew more about?
- The building is meant to bring a positive impact to the community. What, in your opinion, are things we should or should not do with respect to the design and function of the building?

Some common answers indicated a desire to learn more about the history of Muscatine's pearl button industry, the city's architecture and preservation efforts, its place in the Underground Railroad, the work of Alexander Clark, and the various settlers, timber barons, and "industries that made Muscatine the city it is today." One common denominator that stands out was community members' expressed hope that the Stanley Center and its new building be a vehicle for educating on the history of Indigenous peoples of the area. "Most important would be the history of the native people who lived here," wrote one participant. A concern over "the erasure of Muscatine's native history" was also voiced.

In response to the survey on the history of Indigenous peoples in the area, the Stanley Center posed their own question, directed specifically to members of the Meskwaki Nation: "How can settlers be accountable to the people of the lands on which they live?"

The Stanley Center commissioned a work of art that now stands in the front corner of their building, adjacent to where Third and Iowa intersect. "Tree of Accountability," designed by artists Deana Dartt, PhD, of the Coastal Chumash Nation, and Nancy Judd, collaborated to create an installation piece made from metal salvaged from the Musser Library's original guardrails, comprising the "tree." The "leaves" are made of pages from several of the library's decommissioned books. And written on the individual leaves are messages of reflection and pleas of hope — "Care for the Earth like it's your body"; "Respect our religion and culture"; "Learn a true American History"; "Always think about how your decisions will affect generations to come" — that are in response to the Stanley Center's query and are intended to prompt conversations between settlers and Native communities.

"Tree of Accountability" represents a distinctly fascinating convergence of commitments on the part of the Stanley Center, and the artists themselves. The physical work of art is a great example of sustainable, cradle-to-cradle design, in which building products are given new life but in a different form, thus extending the lifecycle of the constituent materials and diverting waste. At the same time, the work acts as a

powerful metaphor for the Stanley Center laying down roots within the land and city that the organization has called home since its founding.

In a prepared statement written by Dartt, delivered by Mark Seaman at the Center's official opening in June 2023, she writes, "We at the Stanley Center are cognizant of our presence in the stolen lands of the Native peoples of the Mississippi delta—the ancient homelands of the groups now known as the Meskwaki, Kickapoo, and Ollagala Sioux. We are committed as an organization and as individuals to leverage our resources in support of Native justice. We have begun a conversation with local Native stakeholders—which is manifest in 'Tree of Accountability'—to begin thinking through what true allyship and meaningful collaboration could mean right here where we live and work. In this way, we move forward from a mere performative 'land acknowledgment' in the direction of true accountability for balance and justice in our shared world."

Part of the Stanley Center's mission is to inform and engage on issues tied to climate action and social justice. In that tradition, educational signage can be found throughout the building, including this one that details the "Tree of Accountability" and the artists who made it.

THE BEAUTY PETAL

Where the Petals Converge

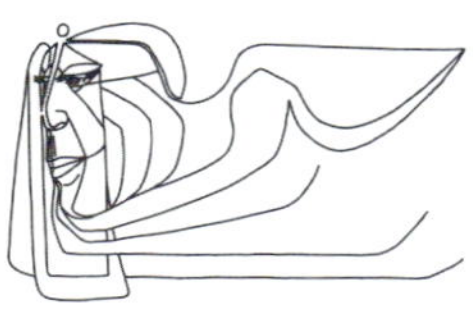

WHERE THE PETALS CONVERGE

Ecosystems thrive when all their components are allowed to function and interact. The more pieces there are the better it is for everything (and everyone) involved.

But whenever native species are removed from the system, or their numbers are diminished, the ecosystem suffers, and its capacity to support life is thus diminished. The same is true for a Living Building. Remove or diminish any Imperative and the associated Petal will not function as designed. Consequently, the building cannot perform to the standards to which it aspires.

This ecosystems analogy and variations on it have been made numerous times when discussing the significance of the Living Building Challenge's seven Petals. They are as interdependent as trees and flowering shrubs are to the birds and bees. But is there a particular Petal that carries more weight than others? Arguably, no. But Beauty, as its name suggests, does have the ability to stand out.

Beauty is the common denominator. It is the glue that binds the other Petals, lending each one greater focus and giving design equal footing with function and performance.

Early in the process, when the Beauty Petal came up in conversations among Stanley Center staff and board members, it had the surprising consequence of segueing into other impactful topics, according to Catherine Elliott, a member of the Stanley Center's governance. "It created some really interesting conversations that then circled back to Equity, employee retention, [and] our organizational goals. [Beauty] gave us perspective on a different set of tools that we could use to achieve those goals."

When building materials are analyzed and procured for their efficiency, lifecycle benefits, and other sustainable attributes, those very materials must also be seen as integral elements that "nurture the innate human/nature connection" (i.e., biophilia), according to the Beauty Petal's written intent. Such environments are more conducive to worker productivity, performance, and collaboration. More so, they make for healthier and happier places to come to work, which in turn assists with recruitment and employee retention. To varying degrees, Beauty informs everything else.

The Stanley Center's Water Petal, for example, is as much about the organization's handprinting relationship with MCSA and the building's cisterns as it is the winding band of pavers at the front entrance, designed to mimic the path of the site's original creek bed. The Stanley Center's Place Petal is both creating a reference habitat for native plant species and designing a beautiful courtyard that enhances the impacts of those habitats. The Energy Petal is finding the right composition of PV modules to service the building's power needs and leveraging those modules — in the form of the canopies — to create a welcoming entryway and an attractive civic gathering space in the adjacent alleyway. And so on and so forth. Each Petal is more than the sum of its parts, and this fact is largely attributable to what the Beauty Petal brings to the table.

When our relationship to a place is defined by how we interact with it (and with others within it), as opposed to avoiding or circumventing it, then our experiences are enhanced by good (or great) design. When there is no consideration given to the innate and very tangible benefits of biophilic design, or when considerate and beautiful design is relegated to secondary status, then the impacts of the Living Building Challenge are severely diminished.

Fortunately, the Beauty Petal is what in many respects keeps the other Petals honest. Otherwise, the International Living Future Institute and like-minded organizations might as well be in the business of building high performance machines with little regard for how they affect our sense of place, belonging, and wellness. Thankfully, this is not the case.

"[Discussion of the Beauty Petal] created some really interesting conversations that then circled back to Equity, employee retention, [and] our organizational goals. It gave us perspective on a different set of tools that we could use to achieve those goals."

CATHERINE ELLIOTT
Member of Governance, Stanley Center for Peace and Security

"The Biophilia Workshop was like our kickoff. I think as designers, we tend to easily come up with things that we want to see in a project. End goal design aesthetics, that sort of thing. In this project, that workshop really was our inspiration and helped steer us to the right place, rather than us coming in with preconceived notions of what something was going to be."

LYNDLEY KENT
Principal, Neumann Monson Architects

The design team strategically reused furnishings from the Stanley Center's existing office through much of their new space, such as this sitting area seen in the foreground

"The unity of vision and the ideas the workshop brought to the team made for a much stronger design. You need to do those kinds of workshops early. Your project won't be as successful if it's left until later in the design process."

SALLY OBERNOLTE
Principal, Neumann Monson Architects

TRANSFORMATION THROUGH NATURE

The integrated design team at Neumann Monson confides that they approached the project with "no preconceived notions" about what the building would look like or how it would feel. Instead, they trusted the process, which in this case started and ended with the two-day Biophilia Workshop, held in September of 2020.

"The unity of vision and the ideas that it brought to the team made for a much stronger design," says Sally Obernolte. "You need to do those kinds of workshops early on. Your project won't be as successful if it's left until later in the design process."

This meeting of the minds (also discussed in detail in this book's Place Petal chapters) doubled as a training of the minds. To the uninitiated, biophilia can be misunderstood as something little more than introducing houseplants and bio-based materials to a space. But at its heart, biophilia is a mandate to transform spaces through nature. That is, to craft built environments with the use of light and space, natural shapes and forms, and all variety of organic elements that are context-driven and honor the building's place and ecology.

Many of these elements are easy to spot. The building's exterior cladding system of charred Accoya is mesmerizing when it catches the sunlight, and its vertically oriented bands lend height and a quiet gravitas to an otherwise stout, low-rise structure. Further drawing the eye, the canopies of bi-facial PV modules that extend over the front and side of the building provide none-too-subtle reminders of how the Stanley Center is harnessing nature's renewable power and doing so with an eye for design.

Inside the building, natural forms and other elements are clear and present. Daylight pouring in through the skylights and four Ficus trees ascending upward; the stones under the stairwells; the wood of the countertops and banquettes and conference table; the potted plants and lettuce walls throughout the open concept living room and beyond are all such examples. And then there are the elements that are hidden in plain sight, such as the thousands of upcycled pearl button "blanks" that comprise the backsplash for the kitchen, a lovely and artistic ode to Muscatine's manufacturing history and its unique ecological place along the Mississippi River.

The Stanley Center's material palette, both inside and out, is a simple and elegant composition in contrasts. Light and dark play off each other throughout. The silver industrial sheen of the large cisterns out front play off the dark yet warm façades of Accoya siding. Natural wood and earthen tones play off the steel of the handrails and that of the "Tree of Accountability" sculpture. References to habitat and the mighty river and watershed that has defined this town for generations can be found all over the place. Just look to the courtyard and its thriving agriculture program, or to any one of the hundreds of native plantings along the building's perimeter. Look to the recycled glass countertops in the kitchen. Look to the patterns of the carpets, chosen for their subtle design references to movement — both linear and circuitous — that recall the Stanley Center's place in Muscatine, its place on the world stage, and its continuous mission to foster partnerships and diplomacy in service of global peace.

Still, even the promise of a two-day, in-depth workshop devoted to exploring all things biophilia did little to convince some on the design team, at least at the outset, that the former Musser Library building could be transformed into something beautiful. "There were a lot of question marks when we got started, because almost everyone didn't like how the building looked," says Lyndley Kent. "But this being a [Living Building Challenge] project, we were all trying to keep an open mind, even more than we normally would." The shuttered Musser Library building, its sturdy foundation notwithstanding, was the proverbial albatross. The Biophilia Workshop represented a way forward, but to what end, no one knew. "How do we even do this? What does this look like?" Kent recalls her colleagues' collective thinking. "But coming out of that, we had a lot of things to draw inspiration from."

"We would begin to explore what meaningful reparations could look like."

NANCY JUDD
Artist and Environmental Advocate

"The leaves will provide a way for the public and the native community to interact with one another, and to interact with this work of art. A visitor might even get ideas about what they themselves can do. That they are moved to action is my goal."

DEANA DARTT
Independent Museum Consultant and Curator at Live Oak Consulting

TRANSFORMATION THROUGH HISTORY

Catherine Elliott, a member of the Stanley Center's governance and the great-granddaughter of Max and Betty, has observed how the Stanley Center comes from a place of privilege. Indeed, as an organization of good will that was established from enormous wealth generated by and within the community of Muscatine, it is obliged to make good on that history, which it has certainly done through decades of local philanthropy and educational outreach.

The organization continually celebrates the achievements of its founders and what Max and Betty's work meant to the world, their country, the State of Iowa, and their neighbors and colleagues in Muscatine. And now, the Stanley Center is seizing the opportunity to look further back in time and use its new headquarters as a vehicle for education and reflection.

"Tree of Accountability" is a new sculptural work that holds prominent placement within the Stanley Center, standing in the atrium's southeast corner, which looks out to the building's front entryway and Iowa Avenue just beyond it. The artist team of Deana Dartt, PhD, an independent curator who is Coastal Chumash and Mestiza, and a descendant of the Indigenous people of California, and Nancy Judd, an environmental educator and designer who descends from white European settlers, conceived of a work that inspires quiet retrospection and a call to action. The "Tree" in question is in fact a sculptural tight-knit grouping of four trees that are built from upcycled aluminum that was once part of the stairway railings in the former Musser Library. Their green leaves are made from the pages of decommissioned library books and are cut to resemble the leaves of the American elm. This choice of breed is not incidental.

Throughout the elm's native growing range, which includes Iowa, the tree's wood was a popular and versatile choice among Indigenous peoples for use in roof construction, cordage, baskets, weapons, medicine, and cooking. "Before European settlement, elms were often chosen by Native Americans as so-called council trees, serving as a signpost for significant tribal gatherings," writes botanist Rob Paratley, for the University of Kentucky's Urban Forest Initiative. Further, in his waning years, circa 1838, the renowned Sauk leader Black Hawk maintained a lodge on the banks of the Des Moines River, where a large elm tree stood, "with its spreading branches overhanging the stream, and flowing from its roots was a crystal spring of pure water," according to the Genealogy Trails History Group. "Here during the sultry summer days of that year Black Hawk was wont to repose and dream over the years of his former greatness and the wrongs that his people suffered."

Dartt also notes that the American elm now suffers from Dutch elm disease "which, for obvious reasons, was a symbol of European erasure of native life." Like any tree grove, large or small, people are drawn in. On the individual leaves, Dartt and Judd inscribed messages that are equal parts hope and caution. "Learn a true American History" reads one; "Support Native

college scholarships" reads another. There are common themes throughout the work, to be certain, but anyone expecting a collection of proclamations and nothing more will be pleasantly surprised.

"Always think about how your decisions will affect generations to come"; "There are over 500 different tribes, with different languages and cultures"; "Educate yourself on Native realities and potentials." There are hundreds more. Each message is an opportunity for reflection and healing. But more so, as the title suggests, and particularly for the majority of people who will have occasion to regularly encounter and engage with the work, it is an opportunity to reflect and accept some semblance of accountability for the horrors that white settlers brought upon the Indigenous peoples of this land.

"Tree of Accountability" is undoubtedly a beautiful work of art. The physical gesture it strikes is calming, a faithful mimic of nature's elegance. Further, the fact that it's composed entirely of upcycled materials is a testament to the building and organization that houses the work, and to the high sustainable performance standard to which the Stanley Center aspires. But the work's beauty also stems from the tragedy and ugliness of history. It is indeed a remarkable artistic accomplishment to forge such beauty under the symbolic and literal weight of generations of betrayal, displacement, and genocide.

"The DNA of the Kiikaapoi, Sauk and Meskwaki, and the Očhéthi Šakówiŋ is literally in the soil on which our living building rests and, therefore, also in the trees which surround it," reads a passage on the Stanley Center's website devoted to

"Peace will not evolve in our time. It must be actively waged!"

MAX STANLEY (1904-1984)
Co-founder, Stanley Center for Peace Security

the work. "We hope to move toward healing the past and forging ahead in partnership with Native people—not simply through acknowledgment but through accountability."

In addition to this pivotal piece, and other featured works of art by local and Indigenous artists, the Stanley Center maintains close ties with The Elizabeth Foundation for the Arts (EFA), a New York City-based non-profit that supports emerging artists, founded in 1992 by Max's and Betty's daughter Jane Stanley-Stephenson, who remains its executive director. (She named it after her mother.) "When deciding the sources of artwork for the building, the opportunity to partner with EFA felt obvious and meaningful," says Mark Seaman. The Stanley Center worked with the foundation to identify eight works by seven artists, whose respective works speak "to a part of our mission: climate change, violence prevention, peace, and justice," Seaman says.

The Stanley Center's new headquarters is a living embodiment of the organization's values and founding mission, curated art collection and all. "To secure peace with freedom and justice" was a tenet of Max Stanley, if not the central tenet of the organization he established with his wife Betty. It was not a mandate he took lightly. He abhorred myopic thinking and revisionist views of history. He viewed human rights as a global concern. "We need to include the most elemental and inalienable human right to which every citizen of this globe is entitled — the freedom from the insecurity of trauma of war, terrorism, and barbarism," Max said. Working to attain that right for *everyone* requires us to not only look forward, but look back as well, and take account for our past sins. Only then, with beauty and grace, do we have the capacity to make things right.

PART IV

A Higher Calling

"This building is role modeling for people that there are places out there where you can work that are beautiful, and value the people who work there. Where you can show up and just be yourself."

KRISTA REGENNITTER
Program Officer, Global Education,
Stanley Center for Peace and Security

"This is a new canvas, and one that largely falls on my shoulders to make more appealing than it already is. It's all very exciting to think about what's certain rooms will become, creating spaces to inspire people. It's fun, but also daunting. It's a big task to take this blank canvas and transform it even more than it already has been, to fit our needs and our vision and our values."

AMY BAKKE
Creative Director, Stanley Center for Peace and Security

"Part of our aspirational statement that we talked about was the ability of this space to attract new talent. That's very important. We want people to want to work here."

MARK SEAMAN
Vice President and Director of Communications,
Stanley Center for Peace and Security

"I love the outdoors. The courtyard is so nice and refreshing. And even inside, with all the plants and our lettuce wall, you have that feeling of bringing nature and the outdoors inside."

PATTY PAPKE
Vice President and Director of Operations,
Stanley Center for Peace and Security

"I would absolutely do this again. It was so worthwhile to end up with a space as lovely as this."

MARK SEAMAN
Vice President and Director of Communications,
Stanley Center for Peace and Security

REFLECTIONS FROM THE STANLEY CENTER

A well-known principle of the modern era is that form follows function; a building's appearance and design should not be in service of itself but arise organically from its intended purpose. But as far as the Living Building Challenge is concerned, neither beauty nor purpose are secondary.

Everything that goes into a Living Building, to varying degrees, is in service of multiple Petals and Imperatives. More so, every tool, tile, fixture, coat of paint, and piece of furniture, among countless other elements, is in service of a higher calling. That calling is to create buildings that give more than they take, that are fully regenerative on both an environmental and community-wide scale, and that are restorative in ways both tangible and intangible.

"For all the effort that went into this building, all the years of design and construction, and the work that went into developing a thorough understanding of its benefits, now that we're here, the feeling is palpable," says Mark Seaman, vice president and director of communications at the Stanley Center. "Being in this space makes you feel differently than being in any other work environment. There's more sustained energy. There's more camaraderie."

The Stanley Center's CEO Keith Porter doubles down on that. "Even before the pandemic we had remote work, and there was very little incentive to come into the office on days when you weren't required. So, more often than not, I'm here, because I really appreciate the building." Porter cites the courtyard becoming a makeshift work and meeting space, and the feeling of community that comes from sitting in the downstairs "living room," with co-workers milling about or stopping for a chat.

Indeed, moments of respite are hard to come by in many professional settings. But the Stanley Center is not most places. Program officer for global education Krista Regennitter calls it "a very peaceful place" that embodies the Center's values and reflects how the organization values its employees. Vice president and director of policy programming strategy Jennifer Smyser highlights how the building "encourages movement" and creates "opportunities for spontaneous or impromptu interactions." She adds:

Pollinators and other wildlife have begun to flock to the Stanley Center's green spaces thanks to the site being re-vegetated with native species of perennials, shrubs, and trees

"Even on the cloudiest of days, there is natural light coming in and touching almost every piece of this space."

The Stanley Center for Peace and Security marked its official opening on June 3, 2023, with a ceremony and reception that included welcome remarks, via pre-recorded video, from Dr. Hilda Heine, former President of the Republic of the Marshall Islands, a strong advocate for climate action, and close ally of the Center. In her remarks, Dr. Heine called the Stanley Center "an excellent example of how living environmentally friendly needs to be scaled up around the globe."

That work obviously begins at the community level, and within Muscatine, the Stanley Center's neighbors and fellow community members have become "enamored" with the new building, according to Dane Lovell, the Center's facilities maintenance superintendent. Since its opening, Lovell has guided dozens of building tours, leading groups of local elementary school students as well as civic leaders and everyone else in between. He has been actively preparing the facility to begin its initial performance cycle, and he manages the site's urban agriculture program as well. In its first growing season, during the spring and summer of 2023, Lovell reports six separate harvests of vegetables, lettuces, herbs, alliums, and edible flowers, most of which were donated to the Muscatine Center for Social Action's (MCSA) food pantry just next door. Smaller quantities were available for staff to take home or for tour groups to engage with and enjoy. "I've never seen kids so excited to eat lettuce off the walls or pull carrots clean out of the ground. From the outside, it doesn't look like there's any place to grow, but things grew fast, and we ended up with more than we anticipated."

The Stanley Center's urban agriculture program will inevitably be adjusted in future seasons to account for crop performance and other factors. As it happens, getting it up and running also required more resource usage than was anticipated. The summer of 2023 in eastern Iowa was very dry, enough to qualify as a drought. Even so, supply chain issues precluded the Stanley Center from fully installing the building's water filtration system, meaning that in the interim they continued to operate using the city's municipal supply, all while using "an excessive amount of water" to keep up with crop production and irrigation needs. Lovell confides, had they begun the building's performance cycle when they originally intended, "we would have not made it through the performance period if we needed to."

Beyond that, the Stanley Center's completed 149kW solar array, inverters, and battery storage room are all "humming," according to Lovell. But it wasn't all systems go upon occupancy. "For the majority of the people on this project, it's their first Living Building. It's a big learning curve for everybody. It's all new to us, too."

"We have really positioned ourselves to educate others and be a guide for clients who want to explore all their options."

DENISE SCHMITZ
Marketing Manager, Neumann Monson Architects

"When we're speaking about materials and their importance, we're now able to emphasize the choice of renovation versus building new, and how sustainability plays a role in all of that."

SALLY OBERNOLTE
Principal, Neumann Monson Architects

"We are expanding our process and exploring ways to replicate the work we did with the Stanley Center on other projects. Working on a project like this during the pandemic also taught us how to collaborate virtually and better communicate across our offices."

KHALID KHAN
Principal, Neumann Monson Architects

REFLECTIONS FROM NEUMANN MONSON

In many respects, Neumann Monson Architects and the Stanley Center are of the same mind. Every end to a project or initiative is simply the beginning of something new; a new goal, a new benchmark, a new set of hurdles to overcome. The goal posts are constantly in motion.

In this vein, the lessons and inspiration that the Neumann Monson team drew from this experience are anything but superficial. “If we hadn’t done this, we probably wouldn’t have pursued the path of certification for our own office renovation,” says Sally Obernolte. In the fall of 2023, the firm completed a renovation of its Iowa City studio, with plans to pursue the Living Building Challenge’s Core Green Building Certification, which the International Living Future Institute describes, in part, as “a simple framework that outlines the 10 best-practice achievements ... to be considered a green or sustainable building.” Noteworthy achievements include dedicating 65 percent of the office to collaborative workspace (up from 43 percent) and reducing the project’s embodied carbon through a comprehensive salvage and reuse program, extending to all ceiling tiles and doors, 99 percent of the glass, and 76 percent of the casework.

“With the Stanley Center, the budget was pretty flexible, and we were able to achieve the full set [of Petals],” says Tim Schroeder. “But if we’re going to be able to apply this to more clients in the future, our thinking was, here’s an opportunity to experiment on ourselves and see how we could achieve something similar on a modest budget. We just dove in. It’s been a learning process for sure.”

There is a restless penchant for innovation among Neumann Monson’s design leaders. They want to explore what is possible and “build new frameworks” for how to design and partner with clients, according to Brandon Fettes. And a big part of doing that involves looking to the recent past. The Biophilia Workshop in particular, held in September 2020, was eye opening. “It had a huge impact on how we exchange ideas,” Fettes says. Looking ahead long term, the firm now considers this kind of early visioning exercise as a malleable yet crucial component of their design process.

There are different forms the workshop can take, depending on the client and nature of the project, according to Lyndley Kent. “We are in the process of figuring out what those key elements are that can work for any project type.” Incidentally, operating in an entirely virtual setting during the early days of the Covid-19 pandemic enabled the design team to create a collaboration platform that was far more transparent and customizable than they were accustomed to. Working with the digital platform Mural, Neumann Monson and its partners were able to build out a comprehensive set of design boards that could evolve in real time.

This inclusive, panoptic design approach, in effect, invited feedback loops and built transparency into the larger design process. “It became really democratic in that sense, because we weren’t focused on just one aspect of the project at a time,” says Khalid Khan. “We could look at the big picture and the small details at the same time, and anybody could chime in.”

It was Khan who referred to “chain reactions” within the Living Building Challenge, and how one design decision will invariably inform and impact hundreds more, from one Petal and Imperative to the next. But those chain reactions are not limited to material choices, energy usage, and performance strategies. They extend to the design professionals and trade experts who are involved in every Living Building project. In the case of the Stanley Center, which represents Neumann Monson’s first formal encounter with the Living Building Challenge, the sequence of exercises and chain link of decisions made will forever inform how the design firm conducts its practice going forward. In the global architecture industry, where innovation tends to move slowly, for a single project to have that kind of impact is nothing short of tremendous.

A LASTING, LIVING IMPACT

This building may have been completed and the ribbon cut, but the lines of communication and collaboration among the project's key stakeholders did not end once the Center's doors opened. The Stanley Center's new headquarters remains a work in progress.

And in response to that collective understanding, monthly QA/QC meetings between the owner, architect, and general contractor still occur to discuss ongoing project closeout tasks, en route to beginning the building's performance period.

As we established in the opening pages of this book, there is no off-the-shelf version of a Living Building one can purchase and build. Context is everything, from a site's ecological conditions to regional economies and every minutia in between. For the Stanley Center to successfully achieve full Living Building certification (an outcome this author has every confidence will come to bear), it will require constant monitoring, tinkering, and optimizing. It will require a continuous evolution of the Center and its values. Fortunately, this high bar has been the standard by which the Stanley Center has operated since the organization was established nearly 70 years ago.

The Living Building Challenge was created in response to a higher calling. So was the Stanley Center for Peace and Security.

When the research for this book began, and the building itself was nearing completion, I posed a question to several members of Stanley Center's board of governance: How do you think Max and Betty Stanley would feel about this endeavor to create a headquarters that is, in the words of Dr. Hilda Heine, "one of the most environmentally friendly and ecologically sustainable buildings in the world?"

According to Warren Heidbreder, "Max and Betty would have said, of course we should do this. Why wouldn't we do this? And what took you so long to come up with this. Get on with it!"

Max and Betty's great-granddaughter, Catherine Elliott, said, "For Max, with his combination of tangible problem-solving skills and engineering background, this building would represent a fascinating intellectual puzzle. From Betty's perspective, she had such a passion for the environment. She thought about composting and restoring a diversity of ecological species; she was really ahead of her time in that sense. I think they'd be thrilled to see what their organization has become in 2023; how it's stepping up to meet new challenges and opportunities. This never would have happened in the 1950s or 1960s. But finding new ways to address persistent problems is what they did, and it's what we're trying to do at all levels in the organization."

At the Center's opening festivities, in June 2023, Jane Stanley-Stephenson remarked to the audience, "So, what would my father say about this? I think he would say, 'You are doing exactly what you must do. Forget the cost, forget the time, forget the rest. You are leading the country. It's good. So, that's what I believe would be his reaction. But let me tell you, before he said it, he would've talked it over with my mom."

The City of Muscatine, the State of Iowa, and the world at large were inarguably made better places because of the Stanleys' namesake organization. That tradition continues with the completion of the Stanley Center's new headquarters. This building will stand as a lasting example of what can be accomplished when brilliant and conscientious minds join forces to create something that is more than just a physical gesture of good will, and instead is a living and breathing gift to humankind. The Stanley Center for Peace and Security is now the embodiment of our highest ideals.

"In the great Stanley tradition, we can't just have a party to celebrate this amazing new headquarters. We all have to do a little work. Stanley Center makes things fun, but they make you work. And the work is to make the world a better place."

ROBERT ORR
Dean of University of Maryland's School of Public Policy, Special Adviser to the UN Secretary-General on Climate Change, delivered at the building's opening and dedication event, June 3, 2023.

AWARDS

AT THE TIME OF PUBLICATION

2024

Masters Award in the $10 - $25 Million Category from Master Builders of Iowa

2023

Best Development Award in Innovative Leadership Category from 1000 Friends of Iowa

American Institute of Architects Iowa Excellence in Design Merit Award

2022

Green Good Design Award by the Chicago Athenaeum: Museum of Architecture and Design and The European Centre for Architecture Art Design and Urban Studies

PHOTOGRAPHS AND ILLUSTRATIONS

Cameron Campbell, Integrated Studios:
F/B Cover, 1, 2, 9, 37, 38, 50, 57, 59, 63, 69, 75, 79, 86, 87, 88, 90, 95, 96, 104, 109, 113, 115, 121, 129, 130, 132 (bottom), 136 (bottom), 141, 142, 146, 147, 150, 153, 155, 156, 161, 163, 166, 169, 171

Neumann Monson:
41(x2), 44, 45, 46, 47, 48(x3), 55, 60, 61, 66, 67, 70 (bottom x4), 71, 73 (right), 74, 77 (left), 91, 92, 95, 98, 101(x2), 102, 106,110, 116, 117, 119 (top), 123, 125, 127, 128 (left), 134, 135, 136 (bottom), 137, 138, 149 (bottom right), 151, 169

Stanley Center:
18, 23, 24(x2), 26, 27(x3), 29x3, 30(x3), 31, 33, 40, 47, 53, 76, 77 (right), 81, 92(x2), 132 (x2), 136 (top x2), 149(x2)

Biohabitats:
84

John Deason:
64 (bottom)

Design Engineers:
99

Environmental Consulting & Technology (ECT):
73 (left)

Joshua Ford: 1
59

ML Johnson Woodworks:
132 (top)

Library of Congress:
15, 70 (top left)

Tony Loconsole:
21

Muscatine Chamber of Commerce:
53 (top and left)

Musser Public Library:
16(x3), 19(x2), 20, 22, 35, 64 (top), 70 (top right)

Darren Pittman:
11, 83

Piper Video Productions:
49

Editta Sherman:
25 (right)

Kent Sissel:
66 (house photo)

Sac and Fox Tribe of the Mississippi in Iowa:
148 (right), 149 (left)

Smithsonian Institute:
148 (left)

Tectonic Video, Inc.:
53 (right bottom), 103

Thrift Books:
25 (left)

University of Iowa, Stanley Museum of Art:
29 (2x artwork)

Four Ficus Benjamina trees are rooted to the Stanley Center's foundation, and have acclimated so well to their new home that they have required pruning to avoid hitting the skylights

MORE BOOKS FROM ECOTONE PUBLISHING

Ecotone publications are available online at the Ecotone Bookstore at **living-future.org/bookstore** and at other select retailers.

All proceeds from book sales go to supporting ILFI advocacy and programming.

THE LIVING BUILDING CHALLENGE SERIES

THE LIVING BUILDING CHALLENGE
by Mary Adam Thomas

BUILDING IN BLOOM
by Mary Adam Thomas

LIVING BUILDING EDUCATION
by Chris Hellstern

GENERATION GREEN
by Michael D. Berrisford

DESERT RAIN HOUSE
by Juliet Grable

THE GREENEST BUILDING
by Mary Adam Thomas

BROCK ENVIRONMENTAL CENTER FOR A LIVING CHESAPEAKE
by Juliet Grable

REGENERATIVE RETROFIT
by Juliet Grable

THE PAE LIVING BUILDING
by Mary Adam Thomas

CHALLENGE & CHANGE
by Mary Adam Thomas

PIKES PEAK SUMMIT VISITOR CENTER
by Juliet Grable

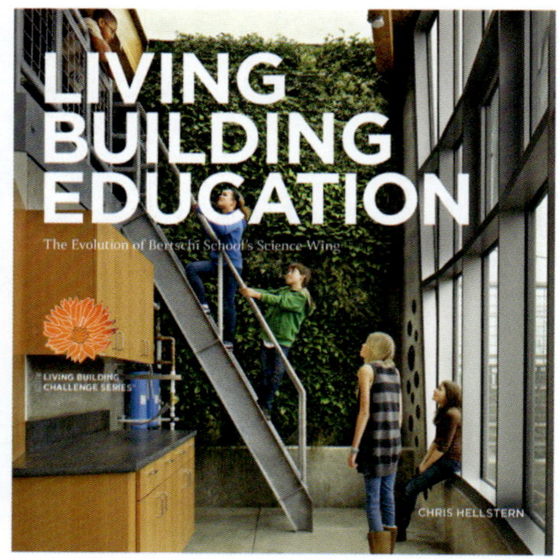

THE POWER OF ZERO:
LEARNING FROM THE WORLD'S LEADING NET ZERO ENERGY BUILDINGS
BRAD LILJEQUIST
WITH JASON F. MCLENNAN BRENDAN COOK JESS CHAMBERLAIN ADAM PAUL AMRHEIN

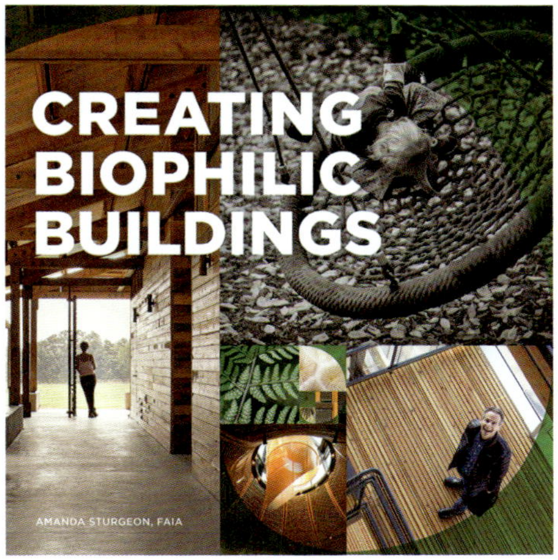
CREATING BIOPHILIC BUILDINGS
AMANDA STURGEON, FAIA

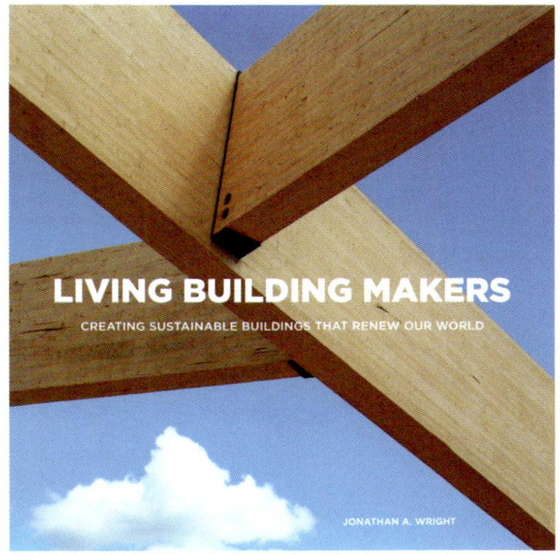
LIVING BUILDING MAKERS
CREATING SUSTAINABLE BUILDINGS THAT RENEW OUR WORLD
JONATHAN A. WRIGHT

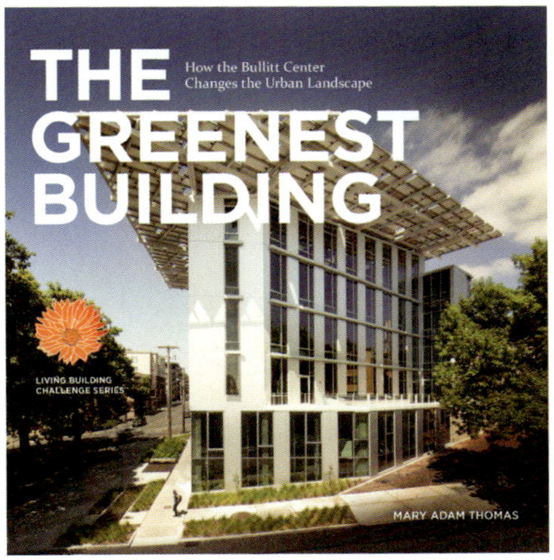
THE GREENEST BUILDING
How the Bullitt Center Changes the Urban Landscape
LIVING BUILDING CHALLENGE SERIES
MARY ADAM THOMAS

BROCK ENVIRONMENTAL CENTER FOR A LIVING CHESAPEAKE
Building a Sustainable Legacy for the Bay
LIVING BUILDING CHALLENGE SERIES
JULIET GRABLE

REGENERATIVE RETROFIT
California's First Living Building
LIVING BUILDING CHALLENGE SERIES
JULIET GRABLE

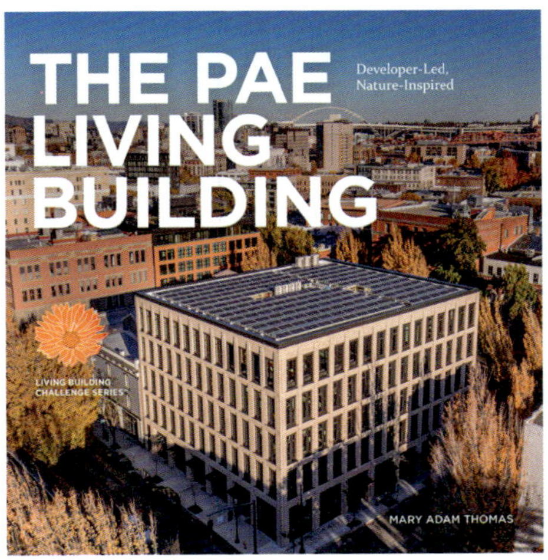
THE PAE LIVING BUILDING
Developer-Led, Nature-Inspired
LIVING BUILDING CHALLENGE SERIES
MARY ADAM THOMAS

CHALLENGE & CHANGE
Miller Hull's Living Building Practice
LIVING BUILDING CHALLENGE SERIES
MARY ADAM THOMAS

PIKES PEAK SUMMIT VISITOR CENTER
A Living Building for America's Mountain
LIVING BUILDING CHALLENGE SERIES
JULIET GRABLE

INTERNATIONAL
LIVING FUTURE
INSTITUTE℠

INTERNATIONAL LIVING FUTURE INSTITUTE

The International Living Future Institute's mission is to cultivate a society that is socially just, culturally rich, and ecologically restorative. To do this, we envision a Living Future and show that it works better in practice and policy. With partners, this means aiming for regenerative communities, buildings, and products, rather than incremental improvements, to make progress much faster and in more transformational ways.

Now more than ever, we need a community working towards a common vision to make progress on the biggest challenges of our time: climate change, health, and justice. And the only path to a regenerative future has justice at its core, because if we aim for sustainability without also seeking justice, we sustain injustice.

The Institute has served as an incubator for more than 500 regenerative buildings around the country, showing what's possible. Now it's time to move past individual projects to coalesce around holistic policy reform at the community scale. A regenerative future in the Anthropocene Era is built on the premise that people are part of nature. When we recognize our place within nature, we create solutions that are efficient, resilient, and refined.

LIVING BUILDING CHALLENGE

The Living Building Challenge™ is the built environment's most rigorous performance standard. It calls for the creation of building projects at all scales that operate as cleanly, beautifully, and efficiently as nature's architecture. To be certified under the Challenge, projects must meet a series of ambitious performance requirements, including net zero energy, waste, and water, over a minimum of twelve months of continuous occupancy.

ECOTONE PUBLISHING

Ecotone Publishing is the non-profit publishing arm of the International Living Future Institute and a key component of the organization's Communication strategy for sharing expert information about green building technologies, materials, and design innovations to create a Living Future. As a publisher, Ecotone produces educational case studies, technical knowledge on renewable energy, plus regenerative and biophilic design, and is the leading source of published information about Living Buildings worldwide. Ecotone also offers professional publishing services to help design firms and organizations document and share their stories, lessons learned, industry reports, and design solutions with others who are also seeking to address the climate crisis and have a positive impact on health and equity in their communities.

**Proceeds from publications and reports re-circulate to support and renew the programs, research, education, and advocacy of the ILFI non-profit organization.*